PEPYS' *DIARY* and
the NEW SCIENCE

Frontispiece to *The History of the Royal-Society of London* (1667) by
Thomas Sprat, with the inscription in the copy presented to the Royal
Society. (Courtesy of the Royal Society.)

The statue is of King Charles II. At our left is Lord Brouncker, many-
times President; at our right, Francis Bacon, spiritual father of the
Society.

Pepys' *DIARY* and the NEW SCIENCE

MARJORIE HOPE NICOLSON

THE UNIVERSITY PRESS OF VIRGINIA
CHARLOTTESVILLE

Preface

THE substance of these three studies was given as the forty-eighth in the series of Page-Barbour Lectures at the University of Virginia in April, 1965. I am deeply appreciative of the honor of having been invited to participate in this distinguished series, established in 1907 and given almost annually since that time.

In preparing the papers for the press, I felt free to go into much more detail for the potential reader than could have been possible in lectures and to introduce collateral material from other experiments performed at meetings of the Royal Society in addition to those specifically mentioned by Pepys. Perhaps I should apologize for the amount of attention I have given to minor works of Samuel Butler in the last chapter on "The Wits." In spite of the fact that the term "The Genuine Remains of Samuel Butler" has been repeated since the eighteenth century, disagreement still exists as to how "genuine" some of the "remains" are, and there are very conflicting theories about the dating of the minor works. I have tried to date the works dealing with science on the basis of internal evidence. Whether my dating can be defended is left for other Butler scholars to determine.

With a very few exceptions, I have limited myself to the period of Pepys' *Diary*, 1660–1669. In the last chapter I have violated my self-imposed chronological limitation by introducing material from the Glanvill-Crosse-Stubbe controversy, since Sprat's *The History of the Royal-Society* and the earlier works of Glanvill, which provoked the controversy, appeared during the period of the *Diary*, and there is no question that Pepys would have been well aware of the attacks on the Society, of which he and Joseph Glanvill were fellow members. I have concluded with a discussion of the most amusing Restoration satire on the New Science, Thomas Shadwell's *The Virtuoso*, played in 1676. There is little doubt that that avid playgoer, Samuel Pepys, attended it, even though he no longer kept the *Diary*.

Ralph Straus in *Carriages and Coaches,* from which I quote in the first lecture, said: "Pepys had a knack of knowing just what posterity would desire to know." So it has proved with the New Science. I have attempted to make clear throughout that it was not his election to the Royal Society that caused Pepys' interest in science. That election occurred in February, 1665, a date almost halfway through the *Diary*. Pepys had bought and used his microscopes and much other scientific equipment before he attended a meeting of the Society. His interest in science was symptomatic of a remarkable period in history, the first age that was as "science conscious" as ours has become. Amateur of science though he was, Pepys was only one of many nonscientific members privileged to write "F.R.S." after their names. In the early years of the Royal Society amateurs far outnumbered scientists. There is nothing surprising in the fact that Samuel Pepys, amateur of science, was installed in 1684 as President of the Royal Society of London for Improving Natural Knowledge.

I have come away from this study with a good deal of respect for Pepys, so far as his interest in science was concerned. To be sure, as I have remarked more than once, Pepys would not have made an ideal teacher of general science for freshmen, since his accounts frequently make an experiment more difficult and abstruse than it was. On one occasion—though only on one—he wrote that he did "lack philosophy enough" to understand the experiment he watched. For the most part he proved an alert, interested, intelligent observer and, in spite of characteristic exaggeration on occasion, a shrewd judge of the distinguished men he met at Arundel House and at the taverns where members of the Society gathered for dinner or supper.

As the basis for my quotations from the *Diary*, I have used the edition by Henry B. Wheatley (London, 1897), and also Lord Braybrooke's notes, insofar as Wheatley did not include them. It has been a constant matter of regret to me that I could not wait for the new edition being edited by William Matthews and Robert Latham and to be published by Bell and Sons. I am delighted to learn from Professor Matthews that the first volume is already in press and that there is a possibility that the complete edition may appear within two or three years. This definitive edition will correct many errors and fill in lacunae, of some of which we are aware in the Wheatley edition. I hope I may be fortunate enough to find that Wheatley did not make as many omissions from the scientific passages as from others and that his text in these particular sections is more accurate than it sometimes proves to be. In referring to the *Diary*, I have consistently used dates rather than page and volume numbers, so that the references may readily be found by a reader using another edition.

Apologies are due to Professor Samuel Mintz for my covering in Chapter III the material he used in his article, "The Duchess of Newcastle's Visit to the Royal Society," *Journal of English and Germanic Philology*, 51 (1952), 168–76. I am sure he will understand that my repetition implied no criticism of his article. It would be unfair to readers for any author dealing with Pepys' *Diary* and the New Science to omit Pepys' racy descriptions of the Duchess, before, during, and after her inruption into a masculine scientific society. Since Professor Mintz was my student for several years, he will not object to my reminding him that he probably heard the tale of Pepys and "mad Madge" in my classroom before he studied it for himself.

I am glad to have the opportunity of acknowledging the assistance I received with the preparation of the manuscript from Miss Elizabeth Horton and Mrs. Winifred Donahue and delighted to thank others who have assisted me in research, particularly Miss Judith Sachs and the cooperative library staff of the Institute for Advanced Study, as well as those cordial hosts to Institute members, the staff of the Firestone Library of Princeton University, particularly Miss Julie Hudson and others in Special Collections. I offer sincere thanks to Professor Lester Conner for his detective work in London, well worthy of Baker Street itself. My greatest debt in this study has been to Mr. George Rousseau, Osgood Fellow of Princeton University, who acted as my research assistant in the Firestone Library as well as in several libraries in New York City. I appreciate his acumen and energy in finding materials for me and his accuracy in catching my too abundant errors. Thanks to these and others, I have had more fun in preparing the Page-Barbour Lectures than in any research I have done since—many years ago—I attempted to decode Swift's cryptic puzzles in the scientific satire of the third book of *Gulliver's Travels*.

In conclusion, I shall try to answer in advance a question invariably asked of any lecturer on Pepys—the pronunciation of his name—by quoting anonymous verses recently discovered, in a nineteenth-century volume of reminiscences, by R. Whitely, Esq., custodian of the Pepys Library at Magdalene College, Cambridge, which I ran across in the *Johnsonian Newsletter*, XXIV (December 1964):

> Oh Samuel, who on some folks' lips
> Are designated Samuel Pips,
> While others follow in the steps
> Of those who call you Samuel Peps,
> At Cottenham the proper step is
> To sound the Y and call you Pepp-iss.
> To such ignorance all Magdalene weeps
> Well knowing you are Samuel Peeps.

M. H. N.

Princeton, March 26, 1965

Contents

Illustrations

I

Samuel Pepys, Amateur of Science

S AMUEL PEPYS began the shorthand entries in his private journal on January 1, 1660, and continued them almost daily for nearly ten years. The final entry for May 31, 1669, was in one of his most characteristic moods: "Thence to 'The World's End,' a drinking-house by the Park; and there merry, and so home late." To this, however, he added a codicil:

And thus ends all that I doubt I shall ever be able to do with my own eyes in the keeping of my Journal, I being not able to do it any longer, having done now so long as to undo my eyes almost every time that I take a pen in my hand; and, therefore, whatever comes of it, I must forbear. . . .
    And so I betake myself to that course, which is almost as much as to see myself go into my grave: for which, and all the discomforts that will accompany my being blind, the good God prepare me!

Pepys was not to become totally blind and was far from going to his grave. He had been in his twenty-seventh year when he made the first entry. He made the last when he was thirty-six. He was to live another thirty-four years, dying at seventy. It is well to remember that the Pepys of the *Diary* was a young man, still in his twenties and early thirties.
    The time span of the journal is almost exactly that of the first decade of the Restoration. More important for my purposes, it also coincides closely with the first decade of the life of

the Royal Society of London, the only academy in Europe that can boast an unbroken history of three hundred years from its establishment down to our own time. Although the Royal Society was not formally chartered until 1662, its founders had been meeting with some regularity from at least November, 1660, some of them, indeed, well before that date, one group at Oxford, others in London. The most important predecessor of the Royal Society had been Gresham College, established by the will of Sir Thomas Gresham at the turn of the century. He had left property to be used by the City of London for the establishment of public lectures in the major fields of learning, law, divinity, music, geometry, astronomy. This early experiment in "adult education" had proved highly successful, since many of the Gresham College lecturers were men of much distinction. When the Royal Society—not yet so called—began its organization meetings, several members were Gresham College professors, the one most familiar to us today, Christopher Wren (youngest of all the founding fathers), being known then rather as astronomer and anatomist than as the architect who was to rebuild London from her ashes. Meetings of the Royal Society were held at Gresham College on Wednesdays, following the Gresham lecture on astronomy, until 1667, when, as Pepys noted on January 7, Henry Howard "gives them accommodation to meet in at his house, Arundell House, they being now disturbed at Gresham College."

Not unnaturally the terms "Gresham College" and "Royal Society" were interchangeable in the minds of many, as they were to Pepys, a fact of which I have had occasion to warn unwary students who, to save themselves time, often try to read the *Diary* by index only. They find comparatively few entries under "Royal Society," [1] since Pepys customarily used "Gresham College," with occasional later references to "Arun-

---

[1] It is interesting to notice that, while Pepys largely thought of the new organization as "Gresham College," his fellow diarist, John Evelyn, made a careful distinction between the old and the new societies, always calling the latter "Royal Society," a title that he himself had given the association. See his *Diary*, December 3, 1661.

dell House." On a very few occasions—not so often as some of
his contemporaries—he spoke of "the college of virtuosoes."
The term "virtuoso"[2] had been added to the English vocabulary only a little earlier in the century. Sometimes compli-
mentary, again satiric, it seemed basically to imply a "collector," whether an antiquarian, a dilettante, or a connoisseur
of *objets d'art,* such as statues, coins, inscriptions, pictures. In
the period of the Restoration the "collector" in science was
often satirized under the title "virtuoso," as we shall see in a
later essay, yet Boyle and other members of the Royal Society
showed no hesitation in using the term seriously of themselves
and of each other. One of the characteristics that made Pepys
greatly interested in the New Science was that he too was, by
instinct and his own training, a collector, of pictures, of books,
of the great collection of ballads, which literary scholars today
associate with his name as inevitably as they associate the *Diary.*

Professor Stimson has called her history of the Royal Society,
Scientists and Amateurs.[3] The title is apt. Today the letters
"F.R.S." after a man's name indicate that he has been recognized for scientific achievement. That was far from true in the
early period. Indeed, there is no better proof that the word
"amateurs" is correctly used than the fact that Samuel Pepys,
scientific amateur if ever there was one, was not only a member of the Royal Society and more than once a member of its
Council, but that in 1684 he was installed as President of the
Royal Society of London for Improving Natural Knowledge.

I

Pepys' education had been the usual one of grammar school
and university. From St. Paul's School he entered Trinity College, Cambridge, migrating shortly to Magdalene College,

[2] On the history and connotations of the term "virtuoso," see Walter E. Houghton, Jr., "The English Virtuoso in the Seventeenth Century," *Journal of the
History of Ideas,* III (1942), 51–73, 190–219.

[3] Dorothy Stimson, *Scientists and Amateurs: A History of the Royal Society*
(New York, 1948).

from which he proceeded A.B. in 1653. His formal education
was not very different from that of John Milton, who had
graduated from St. Paul's twenty-five years earlier than Pepys.
As Professor Clark pointed out in connection with Milton,
education in the English grammar schools was as exclusively
literary as it had been in Roman schools of the first century.
"In Imperial Rome and in Renaissance England all seven of
the Liberal Arts were honored as the basis of a liberal education, but in both periods the mathematical arts of the quadrivium (Arithmetic, Geometry, Music, and Astronomy) were
honored more than taught. The core, flesh and skin of the
educational apple were comprised in the linguistic side of the
trivium (Grammar, Logic and Rhetoric)." [4] Pepys apparently
profited by his training in the trivium. He seems to have read
Latin with pleasure throughout the period of the *Diary*. On
June 13, 1662, he noted: "Up by 4 o'clock in the morning, and
read Cicero's Second Oration against Catiline, which pleased
me exceedingly; and more I discern therein than ever I
thought was to be found in him." He often carried a volume of
Latin plays in his pocket, mentioning on one occasion (July 5,
1663), when "an old doting preacher preached," that he read
"in my book of Latin plays, which I kept in my pocket." He
was sufficiently versed in Latin to note critically: "Our navy
chaplain preached a sad sermon, full of nonsense and false
Latin" (April 27, 1662). As we shall later see, he was quite
capable of following a Latin lecture.[5]

In addition to Latin, he read French with ease and kept up
his reading in that language, since his wife was French and
they often read aloud to each other. He seems to have had a
good knowledge of Spanish also, since he occasionally men-

[4] Donald Clark, *Milton at St. Paul's School* (New York, 1948), p. 3.
[5] In 1663 he seemed to feel that his Latin was getting rusty. He had been inquiring into his brother's academic progress and had set him some reading in a
Latin Bible. He noted on June 21, 1663: "Up betimes, and fell to reading my
Latin grammar, which I perceive I have great need of, having lately found it by
my calling Will to the reading of a chapter in Latin, and I am resolved to go
through it."

tioned visiting a bookseller's for the particular purpose of looking over Spanish books.[6]

As the *Diary* shows, Pepys was a voracious reader of all kinds of books, reading late at night and early in the morning by candlelight, to the great detriment of his sight, and noting often, as on March 18, 1668: "So parted, and I to bed, my eyes being very bad; and I know not how in the world to abstain from reading." Visits to booksellers are frequently mentioned, as he began to build up the library of which he was justly proud, which remains one of his memorials. He frequently mentions having his books rebound in fine bindings with special markings and suggests his pride in the "closet" in which they were housed, for which he ordered the "presses"[7] in which they are still shelved at Magdalene College.

Even a casual reader of the *Diary* becomes quickly aware that Pepys' greatest literary enthusiasm was drama. Indeed the *Diary* is one of the most valuable indexes we have to plays performed during the early Restoration period.[8] "Troubled in mind that I cannot bring myself to mind my business, but to be so much in love of plays," Pepys wrote on August 17, 1661, when he had seen Davenant's *The Witts* for the second time in three days. His troubled mind did not prevent his seeing it again within a week, not to mention twice more when it was revived in 1667 and 1669.

However, his tastes in reading were broad, as the catalogue of his library shows. His account of one group of purchases may serve as a suggestion of many. Having saved three pounds on the stationer's bill (money which presumably belonged to His Majesty, yet which Pepys, like most official servants of his time, felt himself privileged to use as a perquisite), he went to his bookseller's in Paul's Churchyard on December 10, 1663,

[6] See, for example, his entries for July 3, 1661, and January 13, 1664.

[7] August 24, 1666: "Comes Sympson to set up my other new presses for my books . . . to my most extraordinary satisfaction; so that I think it will be as noble a closett as any man hath."

[8] Cf. Helen McAfee, *Pepys on the Restoration Stage* (New Haven, 1916); Montague Summers, *The Playhouse of Pepys* (London, 1935).

"calling for twenty books to lay this money out upon." He found himself

at a great losse where to choose. . . . Could not tell whether to lay out my money for books of pleasure, as plays, which my nature was most earnest in; but at last, after seeing Chaucer, Dugdale's History of Paul's, Stow's London, Gesner, History of Trent, besides Shakespeare, Jonson, and Beaumont's plays, I at last chose Dr. Fuller's Worthys, the Cabbala, or Collections of Letters of State, and a little book, Delices de Hollande, with another little book or two, all of good use or serious pleasure.

After his election to the Royal Society, his reading expanded still more, and we find him reading and often buying scientific works, some much more technical than we should expect of an amateur.

The neglect of the quadrivium in his education comes with a shock to modern readers when they first realize that Samuel Pepys did not know simple arithmetic and that he could not multiply or divide. Early in his post as Clerk of the Acts of the Navy he began to realize his great need of arithmetic in the business of his office and decided to do what he could to obtain it. He entered in his journal for July 4, 1662: "Comes Mr. Cooper, mate of the Royall Charter, of whom I intend to learn mathematique, and do begin with him to-day." His tutor at once discovered that he did not know the multiplication tables and set him to learn them. During the next week Pepys mentions almost daily his attempt to master them: "Up at four o'clock and hard at my multiplication. . . . Up at four o'clock, and hard at my multiplication-tables, which I am now almost master of." He seems to have succeeded well enough in his elementary arithmetic that within a short time he was able to put his knowledge to practical use in attempts to discover the extent to which dishonest practices of the day were cheating the King in the purchase of ship's timber by chicanery in methods of measuring.[9]

[9] There are a number of references during this period to attempts on Pepys' part to find out more about the measuring of timber. See particularly the entry for August 18, 1662: "Mr. Deane . . . showed me the whole mystery of

Pepys' lack of arithmetic was by no means unique in the seventeenth century. Contemporaries like Sir Thomas Browne might write in exalted terms of "the mystical mathematics of the City of Heaven" and know a great deal about Pythagoreanism, but this was no assurance that they could add and subtract. Pepys' ignorance had been shared in his youth by no less a person than John Wallis, one of the most important scientists of the Restoration period, the most distinguished English mathematician before Newton, a father of both differential and integral calculus. His letter on the subject of arithmetic has been widely quoted in treatments of education in the seventeenth century. He had never heard of arithmetic until, when he was about fifteen, his brother, who was destined for trade rather than the professions, lent him the textbook he was studying. Fascinated, Wallis hunted for other books of the kind, "for I had none to direct me, what books to read, or in what method to proceed." His next sentence explains the situation which startles us when we learn of Pepys' inability to multiply: "Mathematics (at that time, with us), were scarce looked upon as Academical Studies, but rather Mechanical—as the business of traders, merchants, seamen, carpenters, surveyors of land, and the like." [10] It is significant that Pepys chose

off-square," a phrase Lord Braybrooke in his edition of Pepy's *Diary* considered an error for *half-square.* Braybrooke's long note at this point, which is repeated in the Wheatley editions, indicates some of the methods in use at the time.

[10] "An Account of Some Passages in His Own Life," appended to Thomas Hearne's Preface to Peter Langtoft's *Chronicle* (Hearne, *Works* [London, 1810], III, cxl). Wallis added: "And amongst more than two hundred students (at that time) in our college [Emmanuel], I do not know of any two (perhaps not any) who had more of mathematics than I . . . which was then but little. . . . For the study of mathematics was at that time more cultivated in London than at the universities." Mark Curtis, *Oxford and Cambridge in Transition, 1558–1642* (Oxford, 1959), added, after quoting the letter, pp. 244–45: "From the notes of Brian Twyne, for instance, it is evident that instructors in mathematics could not count on their pupils having any knowledge of Arabic numerals, to say nothing of simple arithmetical operations performed in such notation. The fault lay with the grammar schools, most of which did not teach arithmetic."
One would expect a person like Milton to know less of mathematics than did Pepys, a fellow graduate from St. Paul's School. On the contrary. During his brief period as a schoolmaster Milton taught his students arithmetic, geometry, and trigonometry. He mentioned that during the Horton period he had gone

for his tutor a mate in the navy who had practical need of
arithmetic. Had he found a teacher somewhat more broadly
grounded, he might have discovered earlier than he did various mechanical aids to arithmetic and other branches of mathematics that were being developed in the period. For instance,
on January 16, 1664, the instrument maker brought him "an
Instrument made of a Spiral line very pretty for all questions
in Arithmetic almost." [11]

It seems extraordinary that the most famous short cut to
multiplication of the century should not have come to Pepys'
attention earlier than it did. Napier's bones had been developed and described by John Napier, the inventor of loga-
rithms, in 1617.[12] The instrument was modified and developed during the century in various ways, in order to make the
manipulation of the "bones" or rods more rapid than it had
originally been. Familiar as they seem to have been throughout the century, Pepys apparently first heard of them in 1667
when he noted on September 26: "To my chamber, whither
Jonas Moore comes, and . . . tells me the mighty use of Napier's bones; so that I will have a pair presently."

Even though, or perhaps because, his tutor forced him to
learn the multiplication tables by heart, Pepys felt that his
adult education in arithmetic had been successful, since, a little more than a year later, he in turn became a mathematical
tutor. On October 21, 1663, after a pair of globes had been delivered to him from the maker, he noted: "This evening . . .

---

regularly to London to study mathematics. Cf. Harris Fletcher, *The Intellectual
Development of John Milton* (Urbana, Ill., 1956), I, 355–84.

[11] This was probably one of a number of variants upon a slide rule, credit
for the invention of which is usually given to William Oughtred. Beside the
straight slide rule, he also devised one of a circular form. The invention of a
circular rule is sometimes attributed to Richard Delamain, a mathematical
teacher, who seems to have invented it independently. See Abraham Wolf, *A
History of Science, Technology, and Philosophy in the Sixteenth and Seventeenth
Centuries* (New York, 1935), pp. 559–60.

[12] The original description of Napier's bones appeared in *Rabdologiae, seu
Numerationis per Virgulas libri duo* (Edinburgh, 1617). A description of the
device, with pictures of the original and a later development on rotatable
cylinders, may be found in Wolf, pp. 558–59.

I begun to enter my wife in arithmetique, in order to her studying of the globes, and, I hope, I shall bring her to understand many fine things." His self-satisfaction in having grasped at least the rudiments of the quadrivium echoes in his judgment upon his clerical brother, on October 17, 1666: "I do fear he will never make a good speaker, nor, I fear, any general good scholar; for I do not see that he minds optickes or mathematiques nor anything else that I can find."

Of course Pepys never became a mathematician as he never became a scientist, yet there is no question that he gained a grasp of arithmetic more than sufficient for the purposes of his office, which he managed expertly in such practical respects. His growing interest in and respect for the subject of mathematics in general show clearly in many changing attitudes. In the early period of the *Diary,* January 10, 1662, he mentioned carrying a message from Lord Sandwich to Sir Paul Neile, "about a mathematical request of my Lord's to him." It seems clear from his passing remarks that he did not understand the message he carried and had no interest in finding out more about it. Within the next few years, particularly after meetings of the Royal Society, when Pepys joined other members at the tavern, he became greatly interested in conversations about mathematics and seems to have been an intelligent listener. Toward the end of the *Diary* he noted on May 8, 1668: "I to Brunckei's house, and there sat and talked, I asking many questions in mathematics to my Lord, which he do me the pleasure to satisfy me in." Pepys, to be sure, was associated with Viscount Brouncker—of whom we shall hear more—in other ways than through the Royal Society, yet the fact remains that Lord Brouncker was a brilliant, if sometimes spotty, mathematician, highly regarded by many. Pepys had come a long way since he learned the multiplication tables.

The last evidence we have of Pepys' interest in mathematics lies well beyond the chronological scope of this lecture, since it occurred in 1693, many years after Pepys had ceased keeping his journal. I mention it because it introduces the most famous name in seventeenth-century science, which does not appear in Royal Society records as early as the period of the *Diary*. In November and December, 1693, Pepys wrote three letters to Isaac Newton, with whom he had a pleasant acquaintance, in part through their common membership in the Royal Society, to which Newton was elected in 1671. Pepys raised a question of probability in throwing dice, putting various possible combinations before Newton. Newton replied in three letters. There is something guarded enough in Pepys' inquiries to make a reader wonder whether Pepys was seeking practical tips on the possibility of his winning in gambling. If not, the correspondence may well have been inspired by Pepys' nephew, John Jackson, then a student at Magdalene, who took an active part in studying Newton's replies and seemed to understand them better than did Pepys. That is of much less significance than is the delightful irony that Pepys, who once had not known the multiplication tables, was responsible for eliciting the only statements Newton ever made on the probability calculus, which otherwise he seems deliberately to have ignored.[13] This irony is balanced by another of the paradoxes that make the charm of this period of history: in 1687 Samuel Pepys, amateur of science, as President of the Royal Society, licensed the publication of Newton's *Principia*.

[13] Florence N. David has studied these letters in "Mr. Newton, Mr. Pepys and Dyse: A Historical Note," *Annals of Science*, XIII (1957), 137–47. Miss David gives the general situation succinctly at the beginning of her article: "Mr. Isaac Newton's small contribution on annuities is known. It is usually thought, probably correctly, that he made no other contribution to either the theory or the practice of probability. For a man of his mental calibre this could only have been because the subject did not interest him, for letters show that he was well aware of such theory of gaming as had been developed in France in 1654. The story is told in letters between Pepys and Newton, and between between Pepys and Tollet, an undertone of calculation being supplied by Pepys' nephew, John Jackson, who was then at Magdalene. The letters are interesting in that they show the state of the probability calculus in England in 1693–94 at a time when we know James Bernoulli was teaching and writing at a far more advanced level. They also provoke the query as to why probability theory is the only branch of pure mathematics to which the English have made no contribution of significance."

II

"To Gresham College (where I never was before)," Pepys noted on January 23, 1661, "and saw the manner of the house, and found great company of persons of honour there." In this first reference in the *Diary* to anything associated with the Royal Society, Pepys was probably referring to the older group, since the Royal Society was still in the early organization stage. In the spring of the next year Pepys accompanied Sir William Penn [14] and others on an official visit to Portsmouth. In the company was Dr. Timothy Clarke,[15] a physician of some note at the time, with whom Pepys seems to have struck up an immediate and pleasant friendship, in part because they shared a room during their stay. "The Doctor and I to bed together, calling cozens from his name and my office," wrote Pepys on April 22, 1662, punning on the doctor's name and Pepys' position as Clerk of the Acts of the Navy. On April 28, "the Doctor and I begun philosophy discourse exceeding pleasant. He offers to bring me into the college of virtuosoes, and my Lord Brouncker's acquaintance, and to show me some anatomy, which makes me very glad; and I shall endeavour it when I come to London." The "college of virtuosoes" was the Royal Society, already attaining an enviable position in the minds of many laymen. In spite of Dr. Clarke's suggestion, however, Pepys was not to be elected to membership for nearly

[14] Sir William Penn (Pepys consistently spells him "Pen"), admiral and general at sea, had recently been appointed Commissioner of the Navy. His name appears dozens of times in the *Diary,* usually accompanied by such epithets as *a cunning rogue, a mean rogue, a hypocritical rogue, a coward, a villain,* and *a rascal.*

[15] Clarke (whom Pepys spells "Clerke") was at this time a candidate for the College of Physicians, of which he was soon to become a fellow. In 1660 he had been appointed physician-in-ordinary to the King's household and later was to become one of the King's personal physicians. He was an original fellow of the Royal Society and a member of the first Council. Pepys reports on May 11, 1663: "Mr. Pierce the surgeon . . . tells me that the other day Dr. Clerke and he did dissect two bodies, a man and a woman, before the King, with which the King was highly pleased."

three years, so that his firsthand acquaintance with the virtuosi
occurs almost halfway through the *Diary*. For that reason I
shall attempt to indicate Pepys' interest in science in general
before and after he was privileged to write "F.R.S." after his
name. It was not primarily the Royal Society that developed
an interest latent in Pepys. Rather, he was symptomatic of
growing tendencies of his age which was rapidly becoming science conscious to an extent unparalleled until our own century.

As I have said, Pepys was to prove a virtuoso in various connotations of the term. In the early days of the *Diary*, however,
his interest in collecting was rather in art and *objets d'art* than
in science. He was to become a connoisseur of painting and to
build up an excellent collection of his own. In the years
1660–62 his references to paintings of various sorts are too
numerous to mention.[16] Wherever he went, he noticed paintings, particularly portraits. As soon as he could possibly afford
it, he commissioned a portrait of his wife and later of himself.
During this same period his references to what may broadly be
called "science" are sparse and desultory. He seems to have enjoyed talk about natural history, as when on February 4, 1662,
he made a lengthy entry about a Mr. Templer "discoursing of
the nature of serpents" and on the tarantula. On January 13,
1662, when he was entertaining at dinner three sons of Robert
Honywood, he wrote: "Mr. Peter . . . after dinner did show
us the experiment (which I had heard talk of) of the chymicall
glasses, which break all to dust by breaking off a little small
end; which is a great mystery to me." These "chymicall
glasses" were Prince Rupert's drops (*lacrymae batavicae*),
supposedly introduced into England from Germany by Prince
Rupert, though there were other accounts of their origin.
They were drops, with long slender tails, formed by dropping
hot glass into water. The whole drop burst into pieces with a
loud detonation if any part of the tail was broken, though the

[16] Wheatley does not index these references. Braybrooke lists twenty-six
references to painting, a majority in the first three years of the *Diary*.

thicker portion was resistant to a great deal of force. Popular as well as scientific interest in them was suggested by Samuel Butler in the second part of *Hudibras*,[17] two years after Pepys' entry:

> Honour is like that glassy bubble,
> That finds philosophers such trouble,
> Whose least part cracked, the whole does fly,
> And wits are cracked to find out why.

Perhaps I can best illustrate Pepys' dual interests as a virtuoso by following one word that is recurrent in his vocabulary, "perspective," a term that implied such basically different things in this period that indexes often prove more confusing than helpful. It is a nice irony that one of Pepys' most frequent uses of the word occurs in connection with various visits he made to the home of Thomas Povey, a member of the Tangier Commission with whom Pepys was closely associated in his professional life—not with enthusiasm as we shall learn—who was also to be his sponsor when Pepys was elected to the Royal Society. On January 16, 1665, he first visited Povey's home to find to his surprise that his professional colleague lived far more graciously than Pepys had ever imagined. After a "most excellent and large dinner," he showed his guests over his house. Although Pepys was properly impressed by Povey's stables and wine cellars, his chief attention on this and later visits was upon rooms "beset with delicate pictures, and above all a piece of perspective in his closett in the low parler." [18] This

---

[17] II.ii.385–88. A long and interesting discussion of the drops may be found in the edition of *Hudibras*, ed. Treadway Russell Nash, (London 1859), I, 319–20. He quotes from Rohault's *Physics*, in which the introduction of the drops into England is attributed to an Edward Clarke who "brought it hither from Holland, and which has travelled through all the universities in Europe, where it has raised the curiosity and confounded the reason of the greatest part of the philosophers." The Royal Society was interested in these drops. On November 19, 1662, Robert Hooke performed an experiment of breaking several bubbles, "of which some broke with a brisk noise, others not." On November 26 he brought in a long report, with his conjectures about the scientific reasons for these curious phenomena.

[18] Evelyn noted July 1, 1664: "Went to see Mr. Povey's elegant house in Lincolns-Inn field, where the Perspective in his Court, painted by Streeter,

"piece of perspective," which Pepys mentions several times, remains something of a mystery to me. I do not think it was the one mentioned by Evelyn in the court as painted by Streater. Pepys also mentioned on May 29, 1664, "his perspective upon his wall in his garden." The perspective which charmed Pepys seems to have been in the house and connected with the "closett," what Evelyn called "the inlaying of his closet." On January 26, a week after his first visit, Pepys saw it again and wrote: "Above all things I do the most admire his piece of perspective especially, he opening me the closett door, and there I saw that there is nothing but only a plain picture hung upon the wall."

The indoor perspective—if I may so discriminate—might possibly have been produced by some variety of camera obscura, a delight of the age. Addison was only one of many men of the seventeenth and eighteenth centuries who thought that the prettiest landscape he had ever seen was one "drawn on the walls in a dark room." [19] Whoever invented it, it had been used by such artists as Alberti and Leonardo and scientifically developed by della Porta and Kepler. Henry Wotton described a picture Kepler had shown him, which he had drawn from the outlines of a picture in the camera reflected on paper. He had made it, he told Wotton, "non tanquam pictor, sed tanquam mathematicus." Used by both artist and scientist, the camera obscura played an important part in art and science. Povey's wide acquaintance with artists and scientists may have

---

is indeede excellent, with the Vasas in Imitation of Porphyrie; & fountaine; the Inlaying of his Closet; but, above all his pretty Cellar, & ranging of his Wine-bottles." The perspective in the court was the work of Richard Streater, a very well known painter of the period, famous as a muralist for his work on the roof of the Sheldonian Theatre and All Souls, Oxford, as well as the ceilings at Whitehall. Pepys was much impressed later with his meeting with Streater, which he describes in detail under the date of February 1, 1669. He was introduced to him by Povey.

[19] "Pleasures of the Imagination," *Spectator*, no. 414. I may refer the reader for references here to a section on the *camera obscura* in my *Newton Demands the Muse* (Princeton, N.J., 1946; republished Hamden, Conn., 1963), pp. 77–81. The interest of poets and prose writers in the instrument is illustrated there by quotations.

made available to him a more scientific model of the sort Robert Hooke demonstrated to the Royal Society a little later,[20] a combination of reflecting and refracting glasses,

A Contrivance to make the Picture of any thing appear on a Wall, Cup-board, or within a Picture-frame, in the midst of a Light room in the Day-time, or in the Night-time in any room that is inlightened with a considerable number of Candles.

A spectator not versed in optics, Hooke declared, seeing the various "Apparitions and Disappearances . . . Motions, Changes and Actions" produced by such means "would readily believe them to be super-natural and miraculous."

If not by the camera obscura, Povey's perspective may have been "all done with mirrors," Pepys having mentioned that when Povey showed him the "unadorned closett" there was a single picture on the wall. The effect could have been produced by such devices as had impressed Evelyn on November 17, 1644, among the many "deceits," which amused or startled visitors in the Borghese Palace: "The perspective is also considerable, composed by the position of looking-glasses, which render a strange multiplication of things resembling divers most richly furnished rooms." Some of these tricks were undoubtedly based upon Euclid's optical and catoptrical theories. That some form of optical illusion was intended is suggested in Pepys' entry on September 21, 1664: "I was afresh delighted with Mr. Povy's house and pictures of perspective, being strange things to think how they do delude one's eyes, that methinks it would make a man doubtful of swearing that ever he saw anything."

"Delusions of the sight" were familiar in both art and demi-science—and, as Hooke's passage shows and we shall again see, science itself was not above using surprising and startling effects. On his travels Evelyn frequently paused to describe such examples of verisimilitude as a painting at the Citronière (February 27, 1644) in which "the sky and the hills . . . are

---

[20] *Phil. Trans.*, III (August 17, 1668), 741–43.

so natural that swallows and other birds, thinking to fly
through, have dashed themselves to pieces against the wall." [21]
Art and science came close together in this period in a common delight in all manner of optical illusions and delusions.
Bacon's Father of Salomon's House in *The New Atlantis* had
"houses of deceits of the senses, where we represent all manner
of feats of juggling, false apparitions, impostures, and illusions" in order to teach their fallacies. Accounts of some seventeenth-century fairs read as if they had been laid in Coney Is-
land. The world did not wait until P. T. Barnum for proof
that all of the people love to be fooled some of the time. The
title of one of John Wilkins' most important works is significant of the temper of those times: *Mathematicall Magick; or,
The Wonders That May Be Performed by Mechanicall Geometry*. The border line between the old and the new "Magick"
was slight in a period in which the wonders of the New Science
might well seem either miracle or the work of a Mephistophe-
les summoned by another Faust. Pepys was by no means
unique in his fascination with Povey's perspectives, which in
part led him to take great interest in various kinds of new optical and other instruments, as we shall see as we continue to
follow his interest in various "perspectives."

Like many men of every generation Pepys had always been
interested in mechanical gadgets. Early in the *Diary* (October
11, 1660), he mentioned stopping in St. James' Park where,
with other "sidewalk superintendents" of his time, he watched
with great pleasure "several engines at work to draw up
water." He was particularly interested in one exhibited by
"Mr. Greatorex," an instrument maker of the day, to whom
we find passing references also in Aubrey's *Lives* and Evelyn's
Diary. A few years earlier (May 8, 1656) Evelyn had seen "an
excellent invention to quench fire," invented by a Greatorex

---

[21] Pepys, too, was delighted with the kind of verisimilitude he might have
found in Apelles' horse or Zeuxis' grapes. On his visit to the King's closet on
October 3, 1660, where he had seen "most incomparable pictures," he particularly stressed a picture of "a book open upon a desk, which I durst have
sworn was a real book."

whom he met when visiting John Wilkins, a more distinguished specialist in mechanical contrivances. During the two years following this first entry Pepys mentions a number of visits to Greatorex' establishment, where he investigated various ingenious devices Greatorex had either invented or developed. On October 24, 1660, for instance, he purchased a drawing pen, examined a wooden chimney jack "that goes with the smoke," and showed particular interest in "lamp glasses, which carry the light a great way, good to read in bed by," one of which he evidently ordered, quite naturally, considering his eyesight. Almost his last reference to Greatorex occurs two years later. On September 22, 1662, he noted: "Walked to Greatorex's . . . and have bespoke a weather glass of him." [22]

As Pepys becomes more and more familiar with the New Science, the earlier, somewhat random enthusiasm for mechanical devices gives way to an increasing interest in more scientific instruments. The name of Greatorex is replaced by those of "Mr. Browne of the Minorys" and "Mr. Spong," both of whom were well known instrument makers, and "Mr. Reeves," the leading glass grinder of the day, whose services were eagerly sought by the Royal Society and who manufactured some of its finest optical instruments. Among the various instruments which Pepys purchased from one or another of them were two other varieties of "perspective," both mechanical aids for drawing, used by some artists and many draftsmen and amateurs for copying. Pepys seems to have liked to draw but perhaps had not sufficient facility to go far without some sort of aid. One instrument, mentioned frequently toward the end

[22] "A very pretty weather-glass for heat and cold" was delivered to him six months later on March 23, 1663, not an unusual delay in a period when instrument workers worked by hand and often without assistants. It is, of course, possible that the disappearance of Greatorex' name from the *Diary* is an indication that he died, though the *DNB* identifies him, somewhat tentatively, with a Ralph Greatorex who lived until 1712. A "Mr. Greatrix," who had developed an engine for going under water, was mentioned by Sir Robert Murray at a meeting of the Royal Society on May 6, 1669 (Thomas Birch, ed., *The History of the Royal Society* [London, 1756], II, 363).

of the *Diary,* was the parallelogram (later called "panto-graph"). On October 27, 1668, "Mr. Spong come, and sat late with me, and first told me of the instrument called parallelo-gram, which I must have one of, shewing me his practice thereon, by a map of England." This was used particularly for copying maps or charts, either in the same size, or larger or smaller, in whatever proportion the copyist desired. It seems to have been a development of a surveying instrument, the "Par-allelogrammum," invented earlier in the century by Robert Fludd and described in his *De Macrocosmi Historia* in 1624.[23] His "Parallelogram," Pepys was careful to note on February 4, 1669, "is not . . . the same as a Protractor, which do so much the more make me value it."

The parallelogram became one of Pepys' enthusiasms. De-livered to him on January 17, 1669, he found it "mighty well, to my full content; but only a little stiff, as being new." This slight defect was promptly adjusted by Spong, who, with other instrument makers, must have vied for Pepys' lucrative cus-tom.[24] So charmed was Pepys with his parallelogram that he bought another, made of brass, which was delivered to him on February 10, 1669. At about the same time Pepys became in-terested in another instrument, which may more truly be called a "perspective."

On February 21, 1666, he asked the advice of Robert Hooke about various aids to drawing. Hooke discussed "the art of

---

[23] Originally Fludd had invented a *baculus geometricus* for taking heights and distances. Descriptions of both instruments, with illustrations, may be found in R. T. Gunther, *Early Science in Oxford,* I (Oxford, 1923–45), 345–47.

[24] Both Reeves and Spong occasionally threw in for good measure a different instrument of one sort or another. We shall see that Reeves added a scotoscope when Pepys purchased a microscope. On the occasion of Pepys' visit about the parallelogram on December 9, Spong showed him "many other pretty things, and did give me a glass bubble, to try the strength of liquors with." Wheatley in his note at this point is quite right, I think, in identifying this with Robert Boyle's hydrometer, "a bubble furnished with a long and slender stem, which was to be put into several liquors to compare and estimate their specific gravity." Wheatley refers to reports of Boyle in *Phil. Trans.,* IV (August 16, 1669), 1001–3, and a later paper in the same journal in June, 1675.

drawing pictures by Prince Rupert's rule and measure, and another of Dr. Wren's; but he says nothing do like squares, or, which is the best in the world, like a darke roome [camera obscura], which pleased me mightily." "Prince Rupert's rule and machine, and another of Dr. Wren's" were perspectographs, several varieties of which had been developed during the century.[25] Some such devices for viewing in correct perspective had been used by Alberti, Leonardo, Dürer, and other artists. A cruder perspectograph had been earlier invented, according to Aubrey, by Francis Potter, of whom we shall hear again, but it was Wren [26] who gave it the great popularity it came to have, which is most amusingly shown in an illustration from *La Dioptrique oculaire* of Père Chérubin d'Orleans (1671), in which we may see up-to-the-minute cherubim, viewing the world and the universe through perspectograph, tele-

[25] Prince Rupert's rule was a perspective to which the Royal Society had devoted a good deal of attention in 1663. On November 11 "Mr. Hooke suggesting, that additions might be made to the invention of Prince Rupert for casting any platform into perspective, so that it might incline and recline, and be fitted to draw likewise solid bodies in perspective, and to describe all kinds of dials, was desired to bring in these aditions in writing, and then to give a description, and to show the practice of the whole. In the meantime, it was ordered, that the Prince's instrument should remain simple, as it was then, without any alteration therein" (R. T. Gunther, *The Life and Works of Robert Hooke*, in *Early Science*, VI [Oxford, 1930], 159). On November 25 "Mr. Hooke brought in an account of his additions to Prince Rupert's perspective engine; and it was ordered, that such an engine should be made for the use of the Society" (*ibid.*, p. 161). On December 16: "Mr. Hooke produced the new perspective engine of Prince Rupert's invention, together with his own additions, to cast embossed things into perspective, as well as platforms. It was ordered, that this engine be showed to Prince Rupert; but that first two rulers of wood be put in the place of the two threads, that directs the parallelism." In *Early Science*, I, 275–77, Gunther discusses and illustrates various "methods for copying or drawing views, and for projecting images of objects upon a screen, either for demonstration or for drawing."

[26] Gunther, *Early Science*, I, 276, describes the instrument: "In Wren's Perspectograph the drawing board was vertical and not transparent, and was arranged laterally in respect of the point of view. The drawing was made by a pencil carried on a rule forming one side of a parallelogram of which two sides are movable (the rule and the upper edge of the chassis keeping the same length). By this means the pencil has transmitted to it all the movements of a pointer, which can be moved by the artist so as to follow the outlines of figures in his field of view."

scope, microscope, and binocular. The perspectograph has
the place of honor, displayed not by one but by two cherubs, at
either end of the canvas.

Earlier in the seventeenth century, after the great interest
aroused by Galileo's first telescopic observations, reported in
1610 in the *Sidereus Nuncius,* the words *perspective* and *pro-
spective* usually implied some variety of telescope.[27] They
continued to have such a connotation throughout the century,
but by the Restoration period telescopes had become so familiar that they were usually called so, rather than by the earlier
synonyms, except among poets and occasional satirists. When
Pepys uses the word *perspective* about an optical instrument,
he refers to a "pocket perspective," as Captain Lemuel Gulliver called it, an early form of binocular, which was used by
many a sea captain and mariner before and after *Gulliver's
Travels.* This was the sort of instrument Pepys first mentioned
on March 23, 1660: "Young Reeve also brought me a little
perspective glass which I bought for my Lord, it cost me 8s."
Much later, on October 23, 1665, on an East Indian ship, Lord
Brouncker had "provided a great dinner, and thither comes
. . . a Perspective glasse maker, of whom we, every one,
bought a pocket glasse."

Perhaps it was this perspective that Pepys irreverently used
in church on May 26, 1667, when an acquaintance insisted
that he occupy a pew in the gallery of the parish church:

Much against my will staid out the whole church . . . but I
did entertain myself with my perspective glass up and down
the church, by which I had the great pleasure of seeing and

[27] See Edward Rosen, *The Naming of the Telescope* (New York, 1947). In
"The 'New Astronomy' and English Imagination," *Science and Imagination*
(Ithaca, N.Y., 1956), p. 39, I wrote of the telescopic nomenclature of the earlier
century: "Its names are many: it is now the 'Mathematicians perspicil' or the
'perplexive glasse' of Ben Jonson; now the 'optick magnifying Glasse' of Donne;
again the 'trunk-spectacle' or 'trunk,' the 'perspective' or 'the glass.' . . . Its
lenses are the 'spectacles with which the stars' man reads 'in smallest characters,' as Butler said in *Hudibras.* Its appearance is frequently commented
upon, as in Davenant's reference in *Gondibert* to 'vast tubes, which like long
cedars mounted lie.' Innumerable figures were coined from it, compliments
paid, insults hurled by its means."

gazing at a great many very fine women; and what with that, and sleeping, I passed away the time till sermon was done.[28]

Like every intelligent layman of the century, Pepys was interested in astronomy, which, with arithmetic, he felt it well to teach his young wife. He wrote on February 15, 1663 (Lord's Day), "After prayers to bed, talking long with my wife and teaching her things in astronomy." On August 7, 1666, Reeves brought him a telescope for trial, but, as so often, the English climate proved unpropitious:

In the evening comes Mr. Reeves, with a twelve-foote glasse, so I left the office and home, where I met Mr. Batelier with my wife. . . . so Reeves and I and they up to the top of the house, and there we endeavoured to see the moon, and Saturne and Jupiter; but the heavens proved cloudy, and so we lost our labour, having taken pains to get things together, in order to the managing of our long glasse. So down to supper and then to bed, Reeves lying at my house, but good discourse I had from him in his own trade, concerning glasses, and so all of us late to bed.

The next evening Pepys arrived home after ten o'clock, tired from a busy day, to find Reeves waiting with two telescopes:

It being a mighty fine bright night, and so upon my leads, though very sleepy, till one in the morning, looking on the moon and Jupiter, with this twelve foote glasse and another of six foote, that he hath brought with him to-night, and the sights mighty pleasant, and one of the glasses I will buy, it being very usefull. So to bed mighty sleepy, but with much pleasure. Reeves lying at my house again; and mighty proud I am (and ought to be thankfull to God Almighty), that I am able to have a spare bed for my friends.

Extravagant as he was about anything he really wanted, Pepys purchased the twelve-foot rather than the six-foot telescope,

[28] Indeed, looking at fair ladies would seem to have been the chief use Pepys made of binoculars. On April 8, 1660, when he was on board a ship that was to assist in the restoration of Charles II, Pepys noted: "The lieutenant and I lay out of his window with his glass, looking at the women who were on board [of two merchantmen they passed], being pretty handsome."

noting on August 19, 1666, that he had seen Jupiter "with my twelve-foote glasse." On August 22 he went home in the evening to find Reeves "and so to look upon the stars, and do like my glasse very well." Pepys settled his bill at once: "did even with him for it, and a little perspective and the Lanthorne that shows tricks, altogether costing me £9, 5s, od."

The fascination of their fathers in the telescope Pepys' generation found in the microscope, which became the most entrancing toy the general public had ever known, a plaything not only of gentlemen but of laides.[29] Pepys naturally shared the general interest which began to sweep over England in the early 1660's. It is interesting to notice that, with only one exception, all his references to the microscope were made before he became a member of the Royal Society. On February 13, 1664:

Creed and I took coach and to Reeves, the perspective glass maker, and there did indeed see very excellent microscopes, which did discover a louse or a mite or sand most perfectly and largely. Being sated with that we went away (yet with a good will were it not for my obligation to have bought one).

He did not make his selection of an instrument until the summer, noting on July 25, 1664, "Thence to Mr. Reeves, it coming just now in my head to buy a microscope, but he was not within." The next day Reeves, who knew a good customer, waited upon Pepys, who went back to the shop where he chose a microscope. On August 7, as he walked home, he met Spong

and he with me as far as the Old Exchange talking of many ingenuous things, musique, and at last of glasses, and I find him still the same ingenuous man that ever he was, and do among other fine things tell me that by his microscope of his owne making he do discover that the wings of a moth is made just as

---

[29] I have discussed the growing enthusiasm for the microscope in "The Microscope and English Imagination," *Science and Imagination,* pp. 155–234. This monograph was originally published in 1935. As I have pointed out, pp. 171–77 and *passim,* there was a whole minor literature on the flea and the louse to which Pepys refers in the next quotation.

the feathers of the wings of a bird, and that most plainly and certainly.[30]

On August 13, 1664, Reeves arrived with the microscope, to which he added a scotoscope,[31] an instrument, as its name indicates, with which to see in the dark. Both name and instrument are now obsolete. The scotoscope was a gift from Reeves. For the microscope Pepys paid £5 10s, "a great price, but a most curious bauble it is, and he says, as good, nay, the best he knows in England, and he makes the best in the world." In anticipation of his microscope Pepys had been reading the first book in which microscopical observations were described in any detail, Henry Power's *Experimental Philosophy in Three Books: Containing New Experiments, Microscopical, Mercurical, Magnetical,* which had just appeared. The first experience of Pepys with the microscope arouses the sympathy of many, such as James Thurber, who remember their bewilderment in freshmen biology:

After dinner up to my chamber and made an end of Dr. Power's booke of the Microscope, very fine and to my content, and then my wife and I with great pleasure, but with great difficulty before we could come to find the manner of seeing any thing by my microscope. At last did with good content, though not so much as I expect when I come to understand it better.

Like everyone who has ever seen it, Pepys was charmed by the finest microscopical book of the seventeenth century, Robert Hooke's *Micrographia: Or Some Physiological Descriptions of Minute Bodies Made by Magnifying Glasses,* [32] which Hooke had hoped to publish earlier but which had been care-

---

[30] One of the most beautiful of the many extraordinary illustrations in Hooke's *Micrographia* (see text below and note 32), which continued to be reproduced for two centuries as the classic example, was of the feathers on a bird's wing.

[31] Apparently this instrument did not prove successful. I can find no description of it. Gunther, *Early Science,* I, 283, merely refers to this passage in Pepys.

[32] London, 1665. This has been republished in Gunther, *Early Science,* Vol. XIII.

fully scrutinized before publication by members of the Society,
to Hooke's immense irritation. Pepys saw it at his bookseller's
on January 2, 1665, and thought it "so very pretty that I presently bespoke it." On January 21 he reported that he had sat
up till two o'clock in the morning "reading of Mr. Hooke's
Microscopicall Observations, the most ingenious book that
ever I read in my life." All these references to the microscope
precede Pepys' membership in the Royal Society, the last by
only a month. There is only one later allusion, on July 29,
1666, which I quote in its entirety since part of it may refer to
the mysterious scotoscope:

So away home to dinner, where Mr. Spong and Reeves dined
with me by invitation. And after dinner to our business of my
microscope to be shown some of the observables of that, and
then down to my office to looke in a darke room with my
glasses and tube, and most excellently things appeared indeed
beyond imagination. This was our worke all the afternoon trying the several glasses and several objects, among others, one of
my plates, where the lines appeared so very plain that it is not
possible to think how plain it was done.

Plaything of the aristocracy and upper-middle-class ladies and
gentlemen though it became, the microscope was to transform
many branches of science, not least medicine, which, within a
few years after the Restoration, became microbiology.

   Almost as if to assist this later author, who has been labori-
ously trying to disentangle the many strands that made up
Pepys' "perspectives," Pepys himself brought nearly all of
them together in one passage, describing how he had spent
"Lord's Day," August 19, 1666:

By and by comes by agreement Mr. Reeves, and after him Mr.
Spong, and all day with them, both before and after dinner,
till ten o'clock at night, upon opticke enquiries, he bringing
me a frame he closes on, to see how the rays of light do cut one
another, and in a darke room with smoake, which is very
pretty. He did also bring a lanthorne with pictures in glasse, to
make strange things appear on a wall, very pretty. We did also
at night see Jupiter and his girdle and satellites, very fine, with

my twelve-foot glasse, but could not Saturne, he being very dark.

Perspectives of science, perspectives of art, perspectives of charming delusion—Pepys spent that Lord's Day very happily. Yet there was much more here than the mere desultory play of a child with toys or a boy with gadgets. Pepys' sincere interest in scientific matters shows in his comments upon one part of his conversation that day:

Spong and I had also several fine discourses upon the globes this afternoon, particularly why the fixed stars do not rise and set at the same hours all the yeare long, which he could not demonstrate, nor I neither the reason of. . . . But it vexed me to understand no more from Reeves and his glasses touching the nature and reason of the several refractions of the several figured glasses, he understanding the acting part, but not one bit the theory, nor can make any body understand it, which is a strange dullness, methinks.

Spong and Reeves were expert technicians, but they were not scientists. Nor was Pepys. Yet unlike the two instrument makers, Pepys possessed the spirit of true inquiry that demands to know why. It was this spirit that made him take such genuine pleasure in the association brought him through the Royal Society with scientists of the first rank, like Robert Hooke, who could always find time to answer the questions of such an intelligent novice as was Samuel Pepys, amateur of science.

III

On February 15, 1665, Pepys entered in the *Diary:* "Thence with Creed to Gresham College, where I had been by Mr. Povy the last week proposed to be admitted a member; and was this day admitted, by signing a book and being taken by the hand by the President, my Lord Brunkard, and some words of admittance said to me." Pepys had earlier seen the charter book he signed that day, in which the names of all fellows were entered. A month before he had gone to Whitehall

to carry on some work with the Duke of York and noted on
January 9, 1665:

Here I saw the Royal Society bring their new book, wherein is
nobly writ their charter and laws, and comes to be signed by
the Duke as a Fellow; and all the Fellows' hands are to be entered there, and lie as a monument; and the King hath put
his with the word Founder.

It seems fitting that the formal introduction of Pepys to the
Royal Society should have come about through two other amateurs, with both of whom Pepys was closely associated in his
professional life as Clerk of the Acts. John Creed was deputy-
treasurer to the Fleet and also secretary of the Tangier Commission, which acted under the Navy Office. Thomas Povey
had been serving as treasurer of the Tangier Commission in
the period immediately preceding Pepys' admission. Neither man had any standing in the world of science, yet both
had been admitted to the Society a little earlier. To many
readers of the *Diary* Povey's sponsorship of Pepys seems ironic.
Pepys admired his house, his perspectives, and his gracious living, but he despised Povey as a colleague, relieving himself constantly of vituperative comments on the handling of affairs on
the Tangier Commission by that "coxcomb," that "puppy,"
"the simple Povy, of all the most ridiculous fool, that ever I
knew to attend to business." On many an occasion there had
been and were to be "hard words between Mr. Povy and I."

"Lord Brunkard," President of the Society, who admitted
Pepys, was also closely associated with him professionally, and
his name is frequent in the *Diary,* apart from his connection
with the Royal Society. When William Viscount Brouncker
had been appointed a Commissioner of the Admiralty in the
December before Pepys' admission, he had come to Pepys for
instructions. Indeed, behind the scenes Pepys acted as mentor
in introducing him to his new duties, and, as we shall see later,
the two men were closely associated during a period when
Pepys had reason to think that he might be dismissed from
office. Brouncker, who held the degree of Doctor of Physick,

was a mathematician of no mean parts—if not one of the greatest, nevertheless a person of real importance in the field,
though his interests were limited and his contributions spo-
radic. He had been appointed President of the Royal Society
in 1663 and continued to serve in that capacity for the longest
term ever known, fourteen years. For some time an able and
devoted officer, he made the mistake of overstaying his welcome, with the result that he was finally forced out. He had
another tie with Pepys in that he too was interested in music,
and indeed his one publication of consequence was a translation of Descartes's *Musicae Compendium* (1653), a book,
Pepys noted, "which I understand not." [33]

For some time, until the pressure of his office became acute,
Pepys attended meetings of the Royal Society regularly, and
even when he was forced to miss the meeting often joined
other fellows at dinner or supper. Of the first meeting he
noted: "But it is a most acceptable thing to hear their discourse, and see their experiments: which were this day upon
the nature of fire, and how it goes out in a place where the air
is not free, and sooner out where the ayre is exhausted, which
they showed by an engine on purpose." We may re-create
those experiments through accounts given in Thomas Birch's
The History of the Royal Society and R. T. Gunther's *The
Life and Work of Robert Hooke*,[34] the latter of which I
quote, since it is the briefer:

Mr. Hooke made an experiment with charcoal enclosed in a
glass, to which nitre being put, and the hole suddenly stopped

[33] On Brouncker see particularly the paper by J. F. Scott and Sir Harold
Hartley in *The Royal Society: Its Origins and Founders,* ed. Sir Harold Hartley (London, 1960), pp. 146–58. This excellent volume of essays, published for
the Tercentenary of the Royal Society, contains new material and is much
more authoritative than most early accounts of the original fellows.

[34] Birch's *History* is not a history in the usual sense of the word. It consists
of four volumes of the minutes of each meeting from the beginning through
1687. I refer to it as either Birch or *History*. Entries from Birch which deal
with Hooke are repeated, sometimes more briefly, in Gunther, *Hooke,* in *Early
Science,* Vols. VI and VII (Oxford, 1930). These volumes also contain letters to
and from Hooke and much other material not included in Birch. The reference
for this entry is VI, 235–36. In later volumes Gunther published Hooke's
Micrographia (Vol. XIII), Hooke's lectures, and other papers (Vols. VIII, X).

again, the fire revived, though no fresh air could get in.

Mr. Boyle affirmed, that gunpowder burns very well in a receiver, out of which the air has been extracted.

He likewise affirmed, that tin mixed with nitre, and Mr. Hooke added, that filings of iron mixed with it, would kindle it. It was ordered, that the experiments should be made.

It was ordered, that Mr. Hooke make trial with a flaming body, and a body heated without flame, whether the heat and flame are preserved best in hot or cold air.

Mr. Hooke made the experiment of gunpowder burning without air.

The formal meeting over, members of the Society, according to their custom, adjourned to the tavern for supper: "After this being done, they to the Crowne Taverne, behind the 'Change, and there my Lord and most of the company to a club supper; Sir P. Neale, Sir R. Murrey, Dr. Clerke, Dr. Whistler, Dr. Goddard, and others of most eminent worth." [35] "Here excellent discourse till ten at night, and then home."

Of his second meeting Pepys noted only: "Thence to Gresham College, where very noble discourse, and thence home busy until past 12 at night." The minutes indicate that on that occasion, too, Hooke had been the chief actor. He had showed a "new small quadrant contrived by himself, to make . . . both celestial and terrestrial observations with more exactness." He had given an account of a dog that had died after the spleen was taken out. Presumably he had also been in charge of three demonstrations, one a continuation of the experiments at the first meeting, on combustion of nitre and sulphur, two others on general problems of the nature of air. Of the third meeting, after a lecture at the older Gresham College, to which I shall return, Pepys noted: "Here was very

[35] Sir Paul Neile was the eldest son of Richard Neile, Archbishop of York. Sir Robert Moray had been one of the founders of the Society and President before the charter was conferred. Daniel Whistler had been professor of geometry at Gresham College; after having been physician to the embassy in Sweden, he returned to England, where he became fellow and later president of the College of Physicians. Dr. Jonathan Goddard, originally professor of Physick at Gresham College, had been physician to Oliver Cromwell. Dr. Timothy Clarke we have already met.

fine discourses and experiments, but I do lacke philosophy enough to understand them, and so cannot remember them." By his word "philosophy," Pepys meant "natural philosophy," or what we today would call "science." The sentence seems curious, since the one demonstration reported by Birch for this meeting was only a variant upon those Pepys had witnessed, with apparently great interest, at the two preceding meetings, an experiment in combustion of nitre and sulphur, while the paper that closed the meeting—which I shall discuss in another connection—was upon the homely subject of the making of French bread, something Pepys or any other layman could have followed readily.

Since the great plague was to forbid all meetings later in this year, and when the Society reconvened a majority of the experiments were on subjects relating to blood transfusion, the subject of my second essay, I shall make no attempt to follow Pepys' visits to the Royal Society chronologically but shall try to indicate the breadth of interests he encountered there by using as point of departure the two greatest scientists of the age, both of whom had been present on the occasion of Pepys' initiation, at which time he had estimated them with shrewd judgment: "Above all, Mr. Boyle to-day was at the meeting, and above him Mr. Hooke, who is the most, and promises the least of any man in the world that ever I saw."

One marked difference between the allusions to the two men is that to Pepys Robert Boyle largely remained an author rather than a person, since during the first eight years of the *Diary*, Boyle maintained his residence in Oxford, and, while he often went to London for Society meetings, Pepys had no such opportunities to meet him casually on the street or in the tavern as he did many other members. As a person, Boyle is mentioned in the *Diary* only twice, in addition to Pepys' comment at the time of his initiation. Pepys had met him much earlier under very different circumstances. As secretary to Sir Edward Montagu (later Earl of Sandwich, the "my Lord" of the *Diary*) Pepys accompanied his superior officer on the expedition made by Montagu to help effect the Restoration by bringing Charles II to Dover. On April 11, 1660, Pepys' office had issued a pass for Robert Boyle to join the expedition, and on April 20 he came on board the *Naseby* on which Pepys was stationed: "This evening came Mr. Boyle on board, for whom I writ an order for a ship to transport him to Flushing. He supped with my Lord, my Lord using him as a person of honour."

Pepys' only other personal allusion occurred many years later, on June 22, 1668, when, knowing that Boyle was attending a meeting of the Council of the Royal Society, he went to Lord Brouncker's: "My business was to meet Mr. Boyle, which I did, and discoursed about my eyes; and he did give me the best advice he could, but refers me to one Turbervill [36] of Salsbury, lately come to town, which I will go to." Apart from that, Pepys' allusions were entirely to Boyle's works, the reading of which on some occasions added to the problem of his eyes. He mentions on April 28, 1667, that he was "mightily pleased with my reading of Boyle's book on colours [37] to-day, only troubled that some part of it, indeed the greatest part, I am not able to understand for want of study." On May 26, 1667, the same day on which he had watched fair ladies through his "perspective" in the Westminster church, he wrote: "I took another book, Mr. Boyle's of Colours, and then read, where I laughed, finding many fine things worthy observation," a statement seemingly inconsistent with the one that preceded and one that followed on June 2. Describing himself as "weary and almost blind with writing and reading so much to-day," he took a boat and went up the river, but instead of resting his eyes, he was "finishing Mr. Boyle's book of Colours, which is so chymical, that I can understand but little of it, but understand enough to see that he is a most excellent man."

Two days later he was reading another of Boyle's books, the

[36] Daubigny Turberville, M.D., came to be considered the leading oculist in London. See Robert L. Pitfield, "A Short Account of Pepys' Oculist, d'Urberville," *Annals of Medical History* (New York, 1928), X, 174–79.
[37] *Experiments and Considerations touching Colours* (London, 1664).

Hydrostatical Paradoxes,[38] in which he took refuge after a quarrel with his wife: "high words between us, but I fell to read a book (Boyle's Hydrostatiques) aloud in my chamber and let her talk." He continued his reading a few days later on June 10: "As long as it was light reading Mr. Boyle's book of Hydrostatics, which is a most excellent book as ever I read, and I will take much pains to understand him through if I can, the doctrine being very useful." This volume continued to be read under odd circumstances. On July 24, 1667, "having tired myself," he indulged in his favorite relaxation of taking a boat, "and so with very much pleasure down to Gravesend, all the way with extraordinary content reading of Boyle's Hydrostatickes, which the more I read and understand, the more I admire, as an excellent piece of philosophy." As he came nearer Gravesend, "we hear the Dutch fleete and our's a-firing their guns most distinctly and loud," a scene reminding the reader of its parallel when Dryden's Eugenius, Crites, Lisideius, and Neander listened from a boat to the cannon of two navies while they debated dramatic poesy. Again in a boat (after a little mild philandering) we find Pepys on August 25, 1667, "reading of Boyle's Hydrostatickes, which are of infinite delight."

On one occasion he read a very different sort of book written by Boyle, more fitting than scientific works for the "Lord's Day," September 15, 1667: "My wife and I to my chamber, where, through the badness of my eyes, she was forced to read to me, which she do very well, and was Mr. Boyle's discourse upon the style of the Scripture,[39] which is a very fine piece, and so to bed."

The last of Boyle's books noted in the *Diary* Pepys bought on April 1, 1668, though he had owned a copy earlier: "Thence called at my bookseller's, and took Mr. Boyle's Book of Formes,[40] newly reprinted, and sent my brother my old

---

[38] *Hydrostatical Paradoxes Made Out by New Experiments* (Oxford, 1666).

[39] *Some Considerations touching the Style of the Holy Scriptures*, which had been published in 1661, reached a fourth edition in 1675.

[40] *The Origin of Forms and Qualities, according to the Corpuscular Philosophy* had originally appeared at Oxford in 1666.

one." Perhaps because of increasing difficulty with his eyes Pepys does not mention this book for six months. On January 29, 1669, he noted: "Home, and there hired my wife to make an end of Boyle's Book of Formes, to-night and to-morrow."

The name of Robert Boyle is too familiar to need further discussion. There is every reason that Robert Hooke's should be equally familiar, as indeed it is coming to be in the history of science. This man was one of the most remarkable and versatile of the seventeenth century, his experiments, inventions, and discoveries spanning every then-known branch of science.[41] Known to modern students chiefly through Hooke's Law (*Ut tensio sic vis*), he is considered by many scholars as having been coequal with Boyle in the development and establishment of Boyle's Law.

Here was one of the truest "sons of Martha" who ever lived, a man of genius who spent most of his life as a drudge, yet who, apart from innumerable other contributions, was as responsible as Sir Christopher Wren for the phoenixlike revival of London from the Great Fire. He had begun his active life at Oxford as a paid assistant to Robert Boyle, a man of wealth. He continued at London as curator of the Royal Society, a position which was paid poorly, when, indeed, it was paid at all.[42] Later he was authorized to find an assistant, but the

[41] Some idea of the extraordinary sweep of his interests is given by E. N. da Andrade in *Royal Society*, ed. Hartley, p. 137:

"Although to many his name is known only through Hooke's Law, outstanding figures in the history of science have been loud in his praises. Thomas Young wrote of the 'inexhaustible but neglected mines of nascent inventions, the works of the great Robert Hook,' a most apt phrase, since Hooke's work contains so much that is suggestive and original, which his restless spirit lacked time to develop. Lalande, writing towards the close of the eighteenth century, said that in Hooke's works were to be found the basic ideas of most modern instruments. Lasswitz, in his standard history of atomism, speaks of Hooke as one of the outstanding figures among the remarkable investigators of his time. Geikie, the geologist, with geology in mind, writes in the highest terms of his versatile intellect and acute observation. The famous astronomers Mandler and J. F. W. Herschel likewise pay high tribute to his remarkable achievement. It is not too much to say that all who have devoted attention to his writings have been struck by the range, the brilliance and the astonishing originality of his thought. It is pleasant to be able to record that after a long period when his admirers were comparatively few, Hooke's genius is now receiving more general recognition." The most recent study is by Margaret 'Espinasse, *Robert Hooke* (Berkeley and Los Angeles, 1962).

[42] In the minutes for October 29, 1666 (Gunther, *Hooke*, VI, 282), Hooke's salary is mentioned as being thirty pounds per annum; although his salary

amount he had to offer was not sufficient to keep an assistant long. "Mr. Hooke was ordered . . . ," "Mr. Hooke was charged . . ."—these are the most reiterated phrases in the minutes of the Royal Society. He was in charge of all the equipment and responsible for furnishing—which in many cases meant inventing and making with his own hands—all instruments necessary for approximately fifty experiments a year. The apparatus required was not always inanimate. Animals, insects, and birds were frequently called for, so that one wonders whether, in addition to all his other duties, Hooke was expected to snare the birds, net the insects, and prowl London streets in search of stray animals, particularly dogs, since the minutes of the Society for May 23, 1667, stated that "the operator was strictly charged to provide dogs from time to time for the use of the society, and of those, who, at the desire of the society, had undertaken to make experiments of several kinds upon them." [43]

Pepys had heard about some of Hooke's experiments even before he himself became a member of the Royal Society. On April 14, 1664, he had walked with Creed to a coffeehouse in Covent Garden where

he told me many fine experiments at Gresham College, and some demonstration that the heat and cold of the weather do rarify and condense the very body of glasse, as in a bolt head with cold water in it put into hot water, shall first by rarifying the glasse make the water sink, and then when the heat comes to the water makes that rise again, and then put into cold water makes the water by condensing the glass to rise, and then when the cold comes to the water makes it sink, which is very pretty and true, he saw it tried.

Pepys, as I shall have reason to remark more than once, would not have made the best of all teachers of general science. It is

---

was badly in arrears, action was delayed from meeting to meeting. On February 14, 1667 (*ibid.,* VI, 292), the arrears were reported as being £114 3*s* 4*d*. The possibility of another curator to assist him was discussed on June 3, 1667, and at various other meetings, On May 30, 1668, Hooke reported that he had found a possible assistant for £20 a year.

[43] *Ibid.,* VI, 305, May 23, 1667.

fortunate that we can always turn to Birch, on this occasion to
learn that Hooke, as we vaguely suspected, was experimenting
on "the stretching and shrinking of glass upon heating and
cooling." We have already heard of Pepys' later conversation
with Hooke about Prince Rupert's and Wren's perspectives
and the camera obscura, as well as Pepys' admiration for
Hooke's *Micrographia*.

Shortly after his admission to the Royal Society, Pepys expressed particular interest in a lecture by Hooke on the comet
of 1664, which had begun to appear in England in November
and could still be observed on March 1, though it was to disap-
pear not long afterward. Pepys had mentioned this comet on
December 15, 1664: "To the Coffee-house, where great talke
of the Comet seen in several places; and among our men at sea,
and by my Lord Sandwich." On Christmas Eve he had seen it
himself.[44] The comet was visible in most parts of the world,
and the Royal Society was to receive many "histories" of it
from the continent and from New England. In his entry for
March 1, 1665, Pepys wrote: "At noone I to dinner at Trinity
House, and thence to Gresham College,[45] where Mr. Hooke
read a second very curious lecture about the late Comet." The
"very curious lecture" contained a hypothesis new to Pepys as
to most of the audience: "Among other things proving very
probably that this is the same Comet, that appeared before in
the year 1618, and that in such a time probably it will appear
again, which is a very new opinion; but it will all be in
print." [46]

[44] "I saw the Comet, which is now, whether worn away or no I know not,
but appears not with a tail, but only is larger and duller than any other star,
and is come to rise betimes, and to make a great arch, and is gone quite to a
new place in the heavens than it was before: but I hope, in a clearer night,
something more will be seen."
[45] Since Hooke's lecture is not mentioned in Birch's minutes, it would seem
that he gave it in his capacity as professor at Gresham College.
[46] I can find no indication that this lecture, which is mentioned elsewhere
as to be published, ever appeared. Hooke's general opinions, however, and a
number of his observations were incorporated into his later paper, *Cometa; or,
Remarks about Comets*, in *Lectiones Cutlerianae; or, a Collection of Lectures
Made before the Royal Society* (London, 1679), pp. 235, 237, 243, and *passim*.

1. Samuel Pepys at the age of thirty-four, by John Hales, 1666. (Courtesy of the National Portrait Gallery, London.)

2. Bust of Mistress Elizabeth St. Michel Pepys in St. Olave's Church, London. (Courtesy of the Reverend Maurice Dean, Rector of St. Olave's, and of the Courtauld Institute of Art.)

Elizabeth Pepys died of a fever on November 10, 1669, aged twenty-nine. In her memory her husband erected this monument, which he could see from his pew in the Navy gallery of the church.

3. Gresham College. (Courtesy of the Society of Antiquaries, London.)

4. Cherubs using perspectographs, telescope, binocular, and microscope. (From Père Chérubin d'Orleans, *La Dioptrique oculaire,* 1671.)

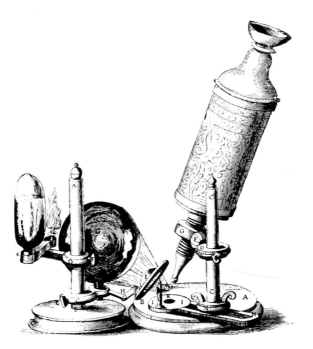

5. Robert Hooke's compound microscope. (From R. T. Gunther, *Early Science in Oxford,* Vol. I, 1923; reproduced by permission of the Executors of the late Dr. R. T. Gunther.)

Here, as so often, Hooke anticipated men with whose names major discoveries are associated, since the theory of periodicity of comets is largely attributed to Edmund Halley, who developed it chiefly on the basis of his close observations of the comet of 1682, "Halley's Comet," the return of which in 1759 he had predicted and which he had traced through records back as early as 240 B.C.[47]

One of the many subjects about which Pepys learned a great deal from Hooke was the nature of sound. On August 8, 1666, he wrote:

Up, and with Reeves walk as far as the Temple, doing some business in my way at my bookseller's and elsewhere, and there parted, and I took coach, having first discoursed with Mr. Hooke a little, whom we met in the streete, about the nature of sounds, and he did make me understand the nature of musicall sounds made by strings, mighty prettily; and told me that having come to a certain number of vibrations proper to make any tone, he is able to tell how many strokes a fly makes with her wings (those flies that hum in their flying) by the note that it answers to in musique during their flying. That, I suppose, is a little too much refined; but his discourse in general of sound was mighty fine.

Even though discussions of sound might prove "a little too much refined," Pepys never seemed to feel that he lacked "philosophy" enough to understand them, because such problems

[47] Halley had begun his work on comets with Newton, who was studying the comet of 1680 with particular relation to his theory of gravitation. In 1664 Hooke could not have gone nearly as far as Halley was to go, since he had, of course, no knowledge of the idea of universal gravitation. Whatever statement he had made in his lecture of 1665 prophesying the return of the comet on the basis of his belief that this was the comet of 1618, in his *Cometa* (which was based chiefly on observations about the comet of 1677) he goes no further than to say that it was "possible" that this was the comet of 1618, "but these are conjectures I shall not insist on" (*op. cit.*, p. 243). The comet of 1618 had attracted great attention, appearing as it did after Galileo had published his earlier telescopic discoveries. Although astronomers made important findings upon it, comets were still associated in the popular mind with dire prognostications. It is interesting to see that Evelyn mentions among his very earliest memories: "I do perfectly remember the great talk and stir about . . . the effects of that comet, 1618, still working in the prodigious revolutions now beginning in Europe, especially in Germany" (1624).

merged with music, always close to Pepys' heart. Hooke does not seem to have been associated with "a new invented instrument to be tried before the College," of which Pepys heard from Oldenburg, Secretary of the Royal Society, on October 5, 1664. Pepys saw and heard the new instrument that same day at a "Musique meeting at the Post-Office." Hooke was undoubtedly present at the demonstration to which "anon come all the Gresham College, and a great deal of noble company." Since Pepys' description is almost the only memorial to "the arched Viall," I shall quote it in its entirety:

> The new instrument was brought called the Arched Viall, where being tuned with lute-strings, and played on with kees like an organ, a piece of parchment is always kept moving; and the strings, which by the kees are pressed down upon it, are grated in imitation of a bow, by the parchment; and so it is intended to resemble several vialls played on with one bow, but so basely and harshly, that it will never do. But after three hours' stay it could not be fixed in tune; and so they were fain to go to some other musique of instruments, which I am grown quite out of love with.

The "Arched Viall," evidently one of many attempts to produce something like the pianoforte, passed into deserved oblivion, from which Pepys' account momentarily rescues it for historians.[48]

Pepys was not at all surprised to see Hooke at a later musical occasion, on February 16, 1667: "Thence away to my Lord Bruncker's, and there was Sir Robert Murray, whom I never

---

[48] Grove's *Dictionary of Music and Musicians,* VII, 974, associates "the Arched Viall" only with Pepys' reference. Evelyn described it the same day, though he gave it no name: "To our Society. There was brought a new-invented instrument of music, being a harpsichord with gut-strings, sounding like a concert of viols with an organ, made vocal by a wheel, and a zone of parchment that rubbed horizontally against the strings."

Mrs. Gretchen Ludke Finney tells me that this was apparently one of many instruments based upon the general principle of the hurdy-gurdy. She called my attention to various such instruments, described and illustrated in the *Encyclopaedia Britannica* and elsewhere, particularly "the Geigenwerk or Geigen-Clavicymbel of Hans Hayden of Nuremburg (c. 1600), a harpsichord in which the strings, instead of being plucked by quills, were set in vibration by small wheels."

understood so well as now by this opportunity of discourse with him, a most excellent man of reason and learning, and understands the doctrine of musique, and everything else I could discourse of, very finely." The recital, attended by "Mr. Hooke, Sir George Ent,[49] Dr. Wren" and many others, was by an Italian group of singers, among whom were two eunuchs. Although the occasion led Pepys to consider various musical matters, including the problem of setting words to music, it did not provoke any basic discussion of the nature of sound, and therefore Pepys reported no part taken in it by Hooke, as he did on a later occasion, on April 2, 1668, when, following a meeting of the Society, he went

with Lord Brouncker and several of them to the King's Head Taverne by Chancery Lane, and there did drink and eat and talk, and, above the rest, I did hear of Mr. Hooke and my Lord an account of the reason of concords and discords in musique, which they say is from the equality of vibrations; but I am not satisfied with it, but will at my leisure think of it more, and see how far that do go to explain it.

Pepys had more than usual reason for his interest on this occasion, since he was at this time working "to the inventing a better theory to musique than hath yet been abroad." [50] I find myself wondering whether Pepys ever discussed with Hooke his own experience at the "Glass-House," where on February 23, 1669, he "had one or two singing-glasses made, which make an echo of the voice, the first that ever I saw; but so thin, that the very breath broke one or two of them." A member of the

[49] Sir George Ent will become more familiar in the second lecture.

[50] I cannot on this occasion enter into various aspects of Pepys' interest in music, except insofar as he mentions discussion of allied problems by scientists of the day. I cannot, however, resist quoting a passage (April 8, 1668) written only a few days after his discussion with Brouncker and Hooke: "So home to my chamber, to be fingering of my Recorder, and getting the scale of musique without book, which I at last see is necessary for a man that would understand musique, as it is now taught to understand, though it be a ridiculous and troublesome way, and I know I shall be able hereafter to show the world a simpler way; but, like the old hypotheses in philosophy, it must be learned, though a man knows a better." So speaks a "New Scientist" about the "Old Philosophy."

Royal Society, at the meeting on November 17, 1670, raised the question whether Hooke "had tried the experiment of breaking a glass with the human voice." Hooke replied that he had tried it, "but found no other success, than that the glass had sounded upon the sound of a man's voice." "Shakespeare and the musical glasses," indeed! We have no record of Hooke's opinion of poetry, but one never caught him napping when it came to anything associated with sound.

The conversation of Pepys, Brouncker, and Hooke about sound was an inevitable corollary to the meeting of the Royal Society which had immediately preceded the tavern talk. "Here, to my great content," Pepys noted:

I did try the use of the Otacousticon, which was only a great glass bottle, broke at the bottom putting the necke to my eare, and there I did plainly hear the dashing of the oares of the boats in the Thames to Arundell gallery window, which, without it, I could not in the least do, and may, I believe, be improved to a great height, which I am mighty glad of.

Here is the immediate ancestor of the ear trumpet, a most important step in the development of artificial aids to hearing. At this meeting, Birch reported,[51]

Mr. Hooke produced a glass receiver for the improvement of hearing. Being tried by holding the neck of it to the ear, it was found, that a stronger sound was conveyed by it, than would have been without it. It was ordered, that at the next meeting there should be brought a better and larger receiver for hearing.

As always, Mr. Hooke was "ordered," and, as nearly always, that indefatigable servant did not only what he was told to do but more. At the meeting on April 9 he produced two "receivers," which again he was ordered to take home and try further, "particularly during the silence of the night, and to bring in an account of their effects." Not only did he slave for them by day, but he listened for them by night, with the result that on April 16 he produced a receiver, "which being tried was found

---

[51] *History,* II, 261.

to make words softly uttered at a distance to be heard distinctly; whereas they could not be so heard without this instrument." [52]

The application of science to human life was Hooke's aim, in spite of the fact that he was basically a theoretical physicist and chemist.

Looking back from our modern perspective, we can see that Robert Hooke was not only the most irreplaceable member of the Royal Society: to a greater extent than they realized, in the early days of its existence he *was* the Royal Society. Even in passing remarks in Pepys' *Diary,* we have some conception of his extraordinary range of interests: we have heard of the nature of air in various experiments on combustion, the principles of sound, the periodicity of comets, the elasticity of glass, the microscopical observations in the *Micrographia* (a book to make its author famous forever), as well as his interest in such lesser matters as perspectographs and the camera obscura. So versatile was he and such a remarkable craftsman that his theoretical work was constantly retarded by the dozens of chores he was forced to perform. His inventive genius would have given him an important place in science, even had his contributions to theory been less than they were. It is seldom, indeed, that the practical and the theoretical have been so extraordinarily combined in one man. When we meet him again, we shall see him not plotting the periodicity of a comet, studying insects under a microscope, or discoursing on air and sound, but setting himself to a homely and practical problem: how to design coaches that they could more easily be turned in narrow London streets. Before that last chapter, I momentarily bid Hooke farewell in words Pepys used of him and John Wilkins—a man second only to Hooke in virtuosity—on May 1, 1665:

[52] Gunther, *Hooke,* VI, 330–31. If Pepys had attended the meeting the previous week—March 26, 1668—he would have found two of Hooke's reports of particular interest, since Hooke "gave a hint of making glasses, by which one might see and read in the dark." He also reported that he had taken the dimensions and drawn the figure of the extraordinary stone found in the body of Sir Thomas Adams, which put Pepys' tennis ball to shame, since "being weighed before the society, was found to weight twenty-two ounces and three-eighths troy weight" (*ibid.,* p. 329).

Deane Wilkins and Mr. Hooke and I walked to Redriffe; and noble discourse all day long did please me, and it being late did take them to my house to drink, and did give them some sweetmeats, and thence sent them with a lanthorn home, two worthy persons as are in England, I think, or the world.

IV

Inevitably, many experiments and discussions of the Royal Society, during the period when Boyle and Hooke were laying the basis for modern physics and chemistry, were theoretical and mathematical, so that an amateur like Pepys might well have lacked "philosophy" enough to understand them. However, the Society did not neglect more practical matters, conscious as the members always were of two recurrent phrases of their spiritual father, Francis Bacon: Experiments of Light and Experiments of Fruit. To Bacon, the first duty of science was to seek for Light: the laws according to which Nature operated. But the scientist must not rest content with Pure Science. He must go on to Experiments of Fruit, the application of Light to the true ends of science, "the benefit and use of man," "the relief of man's estate."

Pepys' *Diary* well shows the interest of the Royal Society in simpler and homelier matters than the weighing of air. Before his election, Pepys mentioned more than once a "little pleasure-boat" that Lord Brouncker and others were building for the King "according to new lines" (August 13, 1662), and he not only saw it racing a "little Dutch Bezan" on September 5, 1662, but even took a trip on it himself (March 2, 1663). Pepys' talk with Hooke about Prince Rupert's and Wren's perspectographs followed a lecture by Hooke about the practical subject of making felt. At his third meeting there was one paper with which Pepys should have had no difficulty. "Among others," he noted "a very particular account of the making of the several sorts of bread in France, which is accounted the best place for bread in the world. So home"—undoubtedly to share his knowledge with Mistress Pepys, with a marked improvement in the breadmaking in the Pepys' kitchen. The chief contribution to this part of the meeting had been a talk by Pepys' fellow diarist (though neither knew the fact about the other) John Evelyn, who delivered a paper called "Panificium: or the several manners of making bread in France . . . where by general consent the best bread is eaten." [53]

Evelyn suggested that "some good English aeconomical person might be consulted for the best of English bread." His paper had led the virtuosi to discuss the possibility of making bread without yeast, or perhaps finding an acceptable substitute for yeast. Of the many practical investigations undertaken by the Society, I shall single out one, improvements in "chariots," a subject in which Pepys became more and more interested as he came closer to that happy day when he could hope to ride in his own coach. [54]

The number of coaches, public and private, increased markedly during the seventeenth century. It is estimated that by 1662 there were approximately twenty-five hundred hackney coaches in London, [55] seriously threatening the living of the watermen, whose barges had once been the most popular form of travel around and near the City. In the early pages of the *Diary* Pepys refers to transportation by boat as a normal part of his life, mentioning particular watermen by name, but as the *Diary* goes on, more frequent references are to coaches. On February 2, 1660, he reported a conversation with "our waterman, White," who "told us how the watermen . . . had lately presented an address of nine or ten thousand hands to stand by

[53] Birch, II, 19. Evelyn does not mention this meeting in his *Diary.*

[54] Ralph Strauss has included most of Pepys' entries, together with some from the records of the Royal Society, in *Carriages and Coaches: Their History and Their Evolution* (London, 1912). He writes, p. 121: "You may learn more of the English seventeenth-century carriages from Pepys than from any other writer; nor is this a matter for wonder. Pepys had a knack of knowing just exactly what posterity would desire to know."

[55] *Ibid.* On p. 125 Straus says: "In this year, 1660, there was a proclamation against the excessive number of hackney-coaches, and two years later Commissioners were appointed for 'reforming the buildings, ways, streets and incumbrances, and regulating the hackney-coaches in the city of London.' Of this body Evelyn was sworn a member in May, 1662."

this Parliament." This, as Pepys was aware, was "a petition against hackney coaches." Six years later, on June 23, 1666, he mentioned in passing that he was forced to change his plans for the day because he "could not get watermen; they being now so scarce."

Sedan chairs had been introduced into England shortly after the death of Elizabeth, though they were never as popular there as on the continent. Ralph Straus in *Carriages and Coaches* mentions [56] an amusing pamphlet of 1636 entitled, "Coach and Sedan Pleasantly Disputing for Place and Precedence, the Brewer's Cart Being Moderator." The same subject was later treated by Swift in his poem, "A Conference between Sir Henry Pierce's Chariot and Mrs. D. Stodford's Chair." Whether for public or private use, the "chariot" became the standard mode of conveyance. It was natural that the Restoration should have been a period during which innovations were introduced into London coaches. Many aristocrats and gentlemen who had followed the King into exile had become familiar with French conveyances. As in other fields, they were undoubtedly more critical of English carriages than their stay-at-home contemporaries might have been. In addition, this was also a period when travelers to the Near and Far East were bringing back news of oriental styles in various fields. It is no surprise to find that on one occasion Hooke was "ordered" to find out more about a Chinese cart about which some members had heard. As usual, he did so and gave a "paper concerning the Chinese cart with one wheel," [57] which proved to be a sort of wheelbarrow. A little earlier [58] Hooke had been instructed to examine an invention brought to the attention of the Society by John Aubrey, "the scheme of a cart, with legs instead of wheels, devised by Mr. Francis Potter," of whom we have heard briefly in passing and whom we shall meet again in connection with blood transfusion. This invention led Samuel Butler to one of his many satiric comments upon the virtuosi.

[56] *Ibid.*, p. 92.   [57] Gunther, *Hooke,* VI, 126.   [58] *Ibid.*, VI, 116.

His Hudibras considered among various means of punishment for an adversary

> to drag him by the heels
> Like Gresham-carts with legs for wheels.[59]

The "new-fashioned chariots" became a reiterated motif in the agenda of the Royal Society, particularly in 1665 and 1666, in the periods immediately before and after the dispersal of the members because of the plague.[60] From entries in the minutes and elsewhere it is clear that they were attempting to make vehicles less cumbersome, smaller and swifter than the former coaches, many of which began to seem more like lumbering stagecoaches than private conveyances. Under Hooke's leadership they were also experimenting on the introduction of springs of various sorts.[61]

The members were particularly interested in improvements in coaches suggested by a "Colonel Blount," who appears in both Evelyn's and Pepys' journals with no fuller name.[62] Eve-

---

[59] *Hudibras*, III, i.1563–64.

[60] The Society had apparently been interested in coaches for some time, since the entry for August 31, 1664, in Gunther, *Hooke*, VI, 192, reads: "It was resolved that the patent for the several new-fashioned chariots be drawn up for Hooke and his assistants."

[61] According to Straus (p. 115), in 1625 an Edward Knapp had received a patent for "hanging the bodies of carriages on springs of steel," but nothing seems to have come of it. The important development in springs came about under the auspices of Hooke and other members of the Society. In connection with the developments going on in coaches during the early Restoration years, I may refer to a letter written on September 13, 1664, by Lady Anne Conway to her husband (in *Conway Letters*, ed. Marjorie Hope Nicolson [New Haven, 1930], p. 229). Lady Conway's preference for the larger coach had its basis in her health: she was a semi-invalid, suffering all her life from an incurable headache. She wrote: "I think it very convenient you should bespeak a coach . . . Whether my Lord Inchequin's coach maker be the best, or others, you will be best able to judge if you take notice of the coaches made by them. That which moved me to fancy him was the commendations he had for making of them easy, but a coach that is not large will never be easy, and therefore I hope you will not dispence with that for the more ease to the horses. It will be convenient also that Mr. Woodshaw take care to see that the seates of the coach be well fitted with the finest Downe, because when there is no special care taken about that, they always stuffe them with such feathers as are very uneasy."

[62] Straus, p. 115, identifies him with Sir Harry Blount, who had traveled extensively, and thinks that his knowledge of coaches was based on his foreign

lyn mentioned him as early as June 11, 1952, as "a great justiciary of the times," whose services he invoked on an occasion when he was robbed. On August 6, 1657, Evelyn saw one of Blount's many inventions, a "waywiser . . . exactly measuring the miles" [63] a vehicle had traveled—apparently the first odometer. On another occasion he had gone with several members of the Royal Society to see Blount's "newly invented Plough" (April 12, 1656). This inventive genius would seem to have been Colonel Thomas Blount, who had been admitted as a member of the Royal Society on February 8, 1665, shortly before the allusions to him in the minutes begin. On March 8 of that year the virtuosi considered a proposal of Blount's for "the improvement of the French chariot, by taking off the burthen from the horse, by means of two small wheels before, retaining the long springy boards." Hooke, practical as always, suggested that if the springs were doubled and shortened, the driver would be more comfortable and the carriage could be more easily turned in narrow London streets. On March 29 Hooke was "desired to take notice of the pole of Prince Rupert's hunting chariot," which he presumably did. On April 12, when Hooke seemed to have made progress on a model of his "chariot with four springs and four wheels," a committee was appointed to "suggest experiments for improving chariots," the membership to include "Sir Robert Moray, Sir William Petty, Dr. Wilkins, Col. Blount, and Mr. Hooke." On April 26 the committee was "desired to meet at Col. Blount's house at Writlemarsh about Col. Blount's model of a char-

---

experiences. Evelyn, however, noted on September 30, 1659, that he had been to visit "Colonel Blount, where I met *Sir Harry, the famous traveller* and water-drinker." The two men were either brothers or cousins, both having their residences at Wrickelmarsh, in the parish of Charlton. Both Evelyn and Pepys mention the fact that Colonel Blount had a vineyard and made his own wine.

[63] When the Royal Society was making preparations for a visit from the King in July, 1663, Christopher Wren suggested among his own devices as an acceptable gift to his Majesty, a "needle that would play in a coach"—evidently a compass—and a "waywiser." See extracts from Wren's letter to Brouncker in Stimson, *Scientists and Amateurs*, p. 78.

iot." [64] At this point Pepys comes into the story, noting on May 1, 1665:

At noon going to the 'Change, I met my Lord Brunkard, Sir Robert Murry, Deane Wilkins, and Mr. Hooke, going by coach to Colonel Blunt's to dinner. So they stopped and took me with them. Landed at the Towerwharf, and thence by water to Greenwich; and there coaches met us; and to his house, a very stately sight for situation and brave plantations; and among others a vineyard, the first that ever I did see. No extraordinary dinner, nor any other entertainment good; but only after dinner to the tryall of some experiments about making of coaches easy. And several we tried; but one did prove mighty easy (not here for me to describe, but the whole body of the coach lies upon one long spring), and we all, one after another, rid in it; and it is very fine and likely to take. These experiments were the intent of their coming, and pretty they are.

The work of the committee, as a committee, was interrupted by the plague, during which the members were dispersed, but Pepys' entries about coaches fill in a few gaps. On September 5, 1665, Pepys had an opportunity to see the new coach Colonel Blount had had made:

Thence home with my Lord Bruncker to dinner where very merry with him and his doxy. After dinner comes Colonell Blunt in his new chariot made with springs; as that was of wicker, wherein a while since we rode at his house. And he hath rode, he says, now this journey, many miles in it with one horse, and out-drives any coach, and out-goes any horse, and so easy, he says. So for curiosity, I went into it to try it, and up the hill to the heath, and over the cart-rutts and found it pretty well, but not so easy as he pretends, and so back again, and took leave of my Lord and drove myself in the chariot to the office.

Pepys mentions going on January 11, 1666, with Lord Brouncker to Gresham College, where they had intended to

---

[64] References appear in both Birch and Gunther, *Hooke,* under the respective dates.

see "a new invented chariott of Dr. Wilkins," but found no
one at home, so they proceeded to Wilkins' home where Pepys
heard "fine discourse . . . to my great joy, so sober and ingenious." He does not, however, give any description of Wilkins' chariot.

When the Society reconvened in January, 1666, the committee on chariots was reactivated. So closely did Pepys come to be
associated with the committee that he might almost have been
a member. He noted on January 22, 1666:

my Lord Bruncker being going with Dr. Wilkins, Mr. Hooke
and others, to Colonel Blunt's, to consider again of the business of chariots, and to try their new invention. Which I saw
here my Lord Bruncker ride in; where the coachman sits
astride upon a pole over the horse, but does not touch the
horse, which is a pretty odde thing; but it seems it is more easy
for the horse, and, as they say, for the man also.

Pepys' description may, as often, leave something to be desired
so far as clarity is concerned, but this time the minutes of the
Society meeting on March 14, 1666, are equally confusing:

Dr. Wilkins and Mr. Hooke [reported] of the business of the
chariots, viz. that after great variety of trial they conceived,
that they had brought it to good issue, the defects found, since
the chariot came to London, being thought easy to remedy. It
was one horse to draw two persons with great ease to the riders,
both him who sits in the chariot, and him who sits over the
horse with a springy saddle; that in plain ground 50-pound
weight, descending from a pulley, would draw this chariot
with two persons. Whence Mr. Hooke inferred, that it was
more easy for a horse to travel with such a draught, than to
carry a single person. That Dr. Wilkins had travelled in it, and
believed, that it would make a very convenient post-chariot.[65]

[65] Birch, II, 66. Straus, p. 116, discussing this invention, quotes, without any
reference, a passage about Hooke's invention which I do not find in Gunther,
Hooke: "Dr. Hooke showed 'two drafts of this model having this circumstantial difference—one of these was contrived so that the boy sitting on a
seat made for him behind the chair and guiding the reins over the top of
it, drives the horse. The other by placing the chair behind and the saddle
on the horse's back being to be borne up by the shafts, that the boy riding
on it and driving the horse should be little or no burden to the horse.' "

The saddle seems to have been entirely Hooke's invention, since the minutes report for April 18, 1666:

The springy saddle contrived by Mr. Hooke, was tried, and an exception being made against the narrowness of the seat, and the way of hanging on the stirrups, it was ordered that against the next meeting it should be made with a full seat, and with the stirrups hanging from the seat itself.

The last entry of any length in the Society minutes, so far as coaches were concerned, appears on May 23, 1666:

Col. Blount and Mr. Hooke were desired to appear on the Saturday following in the afternoon at St. George's Fields, with their new chariots, to compare them together; and it was requested, that as many of the Society as conveniently could, would meet them there.

Unfortunately Pepys was not at that meeting of the Royal Society and on the following Saturday was busily engaged in the work of his office, both morning and afternoon, so that we have no firsthand account of the informal meeting in the park when Blount and Hooke displayed their chariots. But although the Royal Society ceases its references to coaches, Pepys' interest grew as it began to seem possible that he might soon begin to look for a coach of his own. As long ago as March 2, 1662, he had indicated this as one of the two major ambitions of his life. On a Lord's Day morning he lay "talking long in bed with my wife about our frugall life for the time to come, proposing to her what I could and would do if I were worth £2000, that is, be a knight, and keep my coach, which pleased her."

By 1667 Pepys' financial situation was such that "most of our discourse is about our keeping a coach the next year, which pleases my wife mightily; and if I continue as able as now, it will save me money." In the meantime Pepys had been carefully studying various kinds of coaches. So rapidly were changes and improvements being made that a modern reader finds himself quite at home in a world where this year's model of a conveyance is always considered a great advance over last year's model. "Comes Mr. Povey's coach," Pepys noted on October 15, 1665, "and so rode most nobly, in his most pretty and best contrived chariott in the world, with many new conveniences, his never having till now, within a day or two, been yet finished." A little later he was invited to ride in a new light chariot of Sir William Penn's, which Pepys reported as "plain, but pretty and more fashionable in shape than any coaches he hath, and yet do not cost him, harness and all, above £32."

In the meantime the *Diary* alludes on several occasions to "glass coaches," which did not imply the chariot drawn by white horses for Cinderella. During this period some coaches were made with glass windows. Pepys was not the only person of his generation who did not approve this most modern model. Some ladies preferred them, because they could display their charms; others were afraid of them. Pepys wrote on September 23, 1667:

My Lady Ashly's speaking of the bad qualities of glass-coaches; among others, the flying open of the doors upon any great shake: but another was, that my Lady Peterborough being in her glass-coach, with the glass up, and seeing a lady pass by in a coach whom she would salute, the glass was so clear, that she thought it had been open, and so ran her head through the glass, and cut all her forehead.

On July 10, 1668, Pepys himself rode in such a coach: "Thence, in the evening, with my people in a glass hackney-coach to the park, but was ashamed to be seen."

Time will not permit me to discuss in detail the many references in the *Diary* to coaches and coachmaking.[66] In spite of all that he had learned through his fellowship in the Royal Society about the new, lighter, faster coaches, Pepys, like his wife, still liked the dignity of the older, larger models. Indeed, the first time he went so far as to order a specific coach, it was large and heavy. Mr. Povey, whom he asked to go and see it, immediately made him feel that he had made a serious mis-

[66] Straus, pp. 126–32, introduces a majority of Pepys' many references to the coaches he considered and those he finally purchased.

take. "He finds most infinite fault with it," Pepys noted on October 30, 1668, "both as to being out of fashion and heavy, with so good reason that I am mightily glad of his having corrected me in it, and so I do resolve to have one of his build." There were problems about stables and horses, about a coachman and his livery. One by one they were settled, and Pepys rode in his own coach for the first time on December 2, 1668. It was a "mighty pleasure," indeed, for Pepys and his wife to go in their own coach to a play "and makes us appear mighty great, I think, in the world; at least, greater than ever I could, or my friends for me, have once expected; or, I think, than ever any of my family ever yet lived."

It seems eminently fitting that we should say our farewells to Pepys and Mistress Pepys on that last May Day which he recorded in his *Diary*. As so often, he was "up betimes." He put on a summer suit for the first time that year, not his best "of flowered tabby vest, and coloured camelott." With gold lace at his wrists, he would not have felt that he could show himself in the streets or in his office. When he came home to dinner at noon, he found his wife "extraordinary fine, with her flowered tabby gown . . . and she would have me put on my fine suit, which I did." It was May Day, and their coachmen and horses shared the finery of master and mistress. (To be sure, the day was lowering and the Pepyses as often had been having a tiff and were "out of humour," but it was May Day and the coach proved that the Pepyses were quality.) We may say our farewells as they set out for the park:

And so anon we went alone through the town with our new liveries of serge, and the horses' manes and tails tied with red ribbons, and the standards there gilt with varnish, and all clean, and green reines, that people did mightily look upon us; and, the truth is, I did not see any coach more pretty, though more gay, than ours, all the day.

A status symbol then as now—the chariot in which a man rides, Pepys, who had had reason for years to be grateful that he could afford hired hackneys on his duties or pleasures found "the Park full of coaches . . . and, what made it worst, there were so many hackney-coaches as spoiled the sight of the gentlemen's." The future President of the Royal Society of London has come a long way from his early poverty and the learning of the multiplication tables. An aristocrat in all but title, he is like those of his descendants who, when they come to own a Rolls Royce, forget how to drive a Model T.

II

The First Blood Transfusions

T O THE Pope's Head . . . and by and by to an exceeding pretty supper, excellent discourse of all sorts, and indeed they are a set of the finest gentlemen that ever I met with withal," Pepys noted, evidently late in the evening of November 14, 1666, when, although he had been unable to attend a meeting of the Royal Society, he had joined some of the members at the tavern. He continued: "Dr. Croone [1] told me, that, at the meeting at Gresham College to-night . . . there was a pretty experiment of the blood of one dogg let out, till he died, into the body of another on one side, while all his own run out on the other side. The first died upon the place, and the other very well, and likely to do well." So, somewhat confusedly, Pepys reported the first experiment in blood transfusion recorded by the Royal Society, which, so far as he knew, was the first that had ever been attempted. His interest in the surviving dog continued for some time. On November 16 Hooke told him that "the dog which was filled with another dog's blood . . . is very well, and like to be so as ever." Again on November 28 he was glad to hear that the dog "is in perfect good health." Pepys' comment on first hearing of the transfusion was uncon-

[1] Dr. William Croone (1633–84), Professor of Rhetoric at Gresham College and one of the first fellows of the Royal Society, was one of the most active members, whose contributions to various scientific fields were much more important than had been realized until recently. See the article by L. M. Payne, Leonard G. Wilson, and Sir Harold Hartley in *The Royal Society: Its Origins and Founders,* ed. Sir Harold Hartley (London, 1960), pp. 211–20.

sciously prophetic of dual responses this daring experiment was to arouse in the public mind: "This did give occasion to many pretty wishes, as of the blood of a Quaker to be let into an Archbishop, and such like; but, as Dr. Croone says, may, if it takes, be of mighty use to man's health, for the amending of bad blood by borrowing from a better body." Satire on the one hand; on the other, at least a dawning awareness that transfusion between two dogs marked the beginning of an important chapter in the history of medicine.

I

Like most laymen Pepys was greatly interested in doctors and medicine. He had himself been "cut of the stone," [2] a painful and dangerous operation in preanesthesia days with surgery still in its infancy. "This day," he wrote on March 26, 1660, "it is two years since it pleased God that I was cut of the stone at Mrs. Turner's in Salisbury Court. And did resolve while I live to keep it a festival. " Proud of the relic with which the surgeon presented him, he noted on August 20, 1664, "I forthwith to bespeak a case to be made to keep my stone, which I was cut of in, which will cost me 25s." In 1667, when the Lord Treasurer (Thomas Wriothesley, Earl of Southampton) was seriously ill from the same cause, Pepys was waited upon by Sir John Winter, who inquired about Pepys' experience and was shown the stone. On May 3 "to my Lord Treasurer, who continues still very ill. I had taken my stone with me on purpose, and Sir Philip Warwick carried it in to him to see, but was not

---

[2] The "stone," the "pox," and the "ague" were among the commonest diseases of the period. One problem that teased physicians was the origin of stones in the kidney or bladder. An important medical treatment of the subject—a decade later than Pepys' reference—indicates the extent to which analogical thinking by "correspondence" between macrocosm and microcosm still governed some forms of science: Dr. Thomas Shirley, *A Philosophical Essay, Declaring the Probable Causes, Whence Stones Are Produced in the Greater World . . . Being a Prodromus to a Medicinal Tract concerning the Causes and Cures of the Stone in the Kidneys and Bladders of Men* (London, 1672).

in a condition to talk with me about it, poor man." [3] Southampton died on May 16.

It was undoubtedly because of his interest in the affliction from which he had been relieved (though Pepys continued to have some attacks of the stone) that the first extended account we find in the *Diary* of a scientific lecture and demonstration—two years before Pepys was admitted to the Royal Society—had to do with the stone. On February 27, 1663, he went by invitation to "Chyrurgeon's Hall" (Barber-Surgeon's Hall in Monkwell Street) to hear a lecture by Dr. Christopher Terne, fellow of the Royal College of Physicians and one of the original members of the Royal Society. The subject was "the kidneys, ureters, &.," and Pepys found the lecture "very fine." After "a fine dinner and good learned company," Pepys was taken with a small group for a close-up view of the kidneys and ureters which had been used for demonstration, "and Dr. Scarborough, upon my desire and the company's did show very clearly the manner of the disease of the stone and the cutting and all other questions that I could think of," which

[3] This was not the only time Pepys carried his stone to show a fellow sufferer. The first mention of Pepys in Evelyn's *Diary* is in connection with the stone and occurs only a short time after Pepys was forced to lay aside his *Diary*. On July 10, 1669, Evelyn noted: "I went this evening to London, to carry Mr. Pepys to my brother Richard, now exceedingly afflicted with the stone, who had been successfully cut, and carried the stone as big as a tennisball to show him, and encourage his resolution to go through the operation." Pepys' last mention of the stone in his *Diary* shows that he finally heard of a stone even larger than his own. He wrote on March 27, 1668: "This day, at noon, comes Mr. Pelling to me, and shows me the stone cut lately out of Sir Thomas Adams' (the comely old Alderman) body, which is very large indeed, bigger I think than my fist, and weighs about twenty-five ounces; and, which is very miraculous, he never, in all his life, had any fit of it, but lived to a great age without pain." Adams' stone was reported by the Royal Society as weighing 25¾ ounces (Birch, II, 254). In his account of Pepys' medical history in "Pepysiana" (*Diary* IV, 45–49), Wheatley quotes from an address by D'Arcy Power (published in *Lancet*, June 1, 1895), who says that Pepys' operation was performed by "James Pearse, who was afterwards surgeon to Charles II and the Duke of York." This would seem to be the "Dr. Pierce" to whom Pepys frequently alludes. As we shall see presently, he was not the best of technicians. In the passage to be quoted shortly, about Pepys' visit to Surgeon's Hall, he speaks about "poor Dr. Jolly"—presumably dead—as if he associated him with the operation; possibly he had assisted Pearce.

agreed with what Pepys remembered hearing from "poor Dr. Jolly" about his own experience.

The most macabre section of Pepys' report had to do with the body from which the organs had been taken, which Pepys saw in the mortuary. It was of "a lusty fellow, a seaman, that was hanged for a robbery.[4] I did touch the dead body with my bare hand; it felt cold, but methought it was a very unpleasant sight." The fact of the hanging led the physicians and surgeons to a dinner-table discussion which might well have upset a weaker stomach than Pepys' seems to have been. Someone told of "one Dillon, of a great family," who had been hanged "with a silken halter . . . (of his own preparing), not for honour only, but it seems, it being soft and sleek, it do slip close and kills, that is, strangles presently: whereas, a stiff one do not come so close together, and so the party may live the longer before killed." There was cold comfort in the hypothesis on which the doctors agreed: "All the Doctors at table conclude, that there is no pain at all in hanging, for that it do stop the circulation of the blood; and so stops all sense and motion in an instant." Here is the first mention in the *Diary* of a subject about which Pepys was to learn much more when he attended the Royal Society, the circulation of the blood. After a second lecture in the afternoon on the heart and lungs of the hanged man Pepys must have felt that he had had a basic course in anatomy.

At the coffeehouse or tavern Pepys was likely to seek out a physician or a group of doctors because he enjoyed their shop-talk. On November 3, 1663, for example, he wrote:

At noon to the Coffee-house, and there heard a long and most passionate discourse between two doctors of physique, of which one was Dr. Allen, whom I knew at Cambridge,[5] and a

---

[4] The gallows were the customary sources of the corpses used by doctors and medical schools for experiment and teaching. The Royal Society also had official sanction for its use of such corpses.

[5] Thomas Allen, M.D., candidate and later fellow of the College of Physicians, and physician to Bethlehem Hospital, of whom we shall hear in connection with a proposed blood transfusion upon an insane patient at Bedlam.

couple of apothecarys; these [the doctors] maintaining chymistry against their [the apothecaries] Galenicall physique; and the truth is, one of the apothecarys whom they charge most, did speak very prettily, that is, his language and sense good, though perhaps he might not be so knowing a physician as to offer to contest with them.

The subject at issue had been perennial throughout the century. Many physicians and more apothecaries followed the teaching of Galen, the great Greek physician of the second century A.D., that illness was a result of imbalance among the "humours" of the body, to be treated by herbal or vegetable remedies, which would correct the "hot, cold, moist, or dry" from which the patient suffered. Against them was coming to be arrayed the newer school of Paracelsians—sometimes iatrochemists—who insisted that the body was like a test tube, its imbalances to be rectified by proper treatment with chemicals. The apothecaries, as a group, tended to lag behind the physicians, since the pharmacopoeias in which they were trained continued largely to be written in Galenical terms of herbal medicines. Many of these and other basic problems were to be solved in Pepys' own time through the advancement of science, particularly by members of the Royal Society, who were laying the physical and chemical bases on which medicine has proceeded. With his interest in medicine in general, it is not strange that the meetings of the Royal Society Pepys attended most regularly and reported most fully should have been those that had to do with animal or human anatomy rather than with what seemed abstractions of physics. These were to come to a climax when the Royal Society first attempted an extraordinary novelty—which we today accept as a routine form of therapy—blood transfusion.

The practice of blood transfusion was new in the Restoration period, but the theory behind it was very old. Far back in human history, we read of blood baths among the Egyptians, of blood sacrifice, whether of animals or men, of Romans drinking the blood of gladiators. At least as early as the

Pythagoreans blood had been recognized as the basic "humour" by physiologists, who read four humours into the body of man and the body of the world. Inherited, though not basically changed, by the Galenists and the Paracelsians, the blood humour continued to be recognized as the preservative of the body of man, as sap of the plant, or "balme" and "balsam" of the body of the world. But therapeutic treatment by blood, until the late seventeenth century, had been confined chiefly to bloodletting and to attempts to warm or cool this humour by herbs among the Galenists or, among the alchemists, by chemicals.

A *locus classicus* of what may broadly be called "transfusion" is found in the seventh book of Ovid's *Metamorphoses,*[6] the story of Medea and Aeson, father of Jason. When Jason returned triumphant from his "labors," fathers and mothers of his companions brought gifts and burned incense upon the altars. But "Aeson is absent from the rejoicing throng, being now near death and heavy with the weight of years." In answer to Jason's plea Medea used her arts "To renew your father's long span of life." Praying to the gods for "juices by whose aid old age may be renewed and made turn back to the bloom of youth and regain its early years," she prepared in her cauldron a magic brew, far more complex and potent than that of Macbeth's witches. She tested it first upon a dry branch of olive, "which grew green, put forth leaves, and suddenly was loaded with olives." Then she turned to the unconscious body of Aeson:

Medea unsheathed her knife and cut the old man's throat; then, letting the old blood all run out, she filled his veins with her brew. When Aeson had drunk this, in part through his lips and part through the wound, his beard and hair lost their hoary grey and quickly became black again; his leanness vanished, away went the pallor and the look of neglect, the deep wrinkles were filled out with new flesh, his limbs had the

[6] With an English translation by Frank Justus Miller (London and New York, 1916), I, bk. VII, ll. 159–295.

strength of youth. Aeson was filled with wonder, and remembered that this was he forty years ago.

The possibility of rejuvenation of a human being by complete change of blood—such was the theory behind the old legend. But until well down in the seventeenth century any such possibility remained in the realm of magic rather than of science. Centuries after Ovid's imaginary account the hope of rejuvenation seems to have been involved in an operation performed on Pope Innocent VIII in 1492. In a state of coma, he is said to have been given the blood of three young men, all of whom died, as did the Pope. How the blood was administered we do not know. Presumably, as in the past, he received it through the mouth. The theory was old, the belief perennial, but the practice of blood transfusion was impossible before Harvey's discovery of the process of circulation of the blood.[7]

[7] Various histories of medicine and also *Garrison and Morton's Medical Bibliography* (2d ed., London, 1954) mention as the first description of blood transfusion, Giovanni Francisco Colle (1558–1631), *Methodus Facile Parandi Tuta et Nova Medicamento* (Venice, 1628), ch. vii. No copy of this is available in America. Since the book was published in the same year as Harvey's, it is doubtful that Colle knew of Harvey's theory. The situation in the pre-Harveian period seems to have been as discussed by F. J. Cole, "The History of Anatomical Injections," in *Studies in the History and Method of Science,* ed. Charles Singer (Oxford, 1921), pp. 287–90: "Although certain transfusion and intravitam injections had been attempted before the doctrine of the circulation was established in 1628, there can be no question that Harvey's work was the fundamental and determining influence in these early injection experiments. . . . At about this time, the possibility of diverting blood from the vessel of one living animal into those of another was first conceived and practised. The earliest writer to mention transfusion experiments was Magnus Pegel in 1604, and others are Andreas Libavius (1546–1616) in 1615 and Johannes [not to be confused with Giovanni] Colle (ob. 1631) in 1628. It is to be noted that these works were published before Harvey's treatise on the circulation of the blood."

Geoffrey Keynes ("The History of Blood Transfusion, 1628–1914," *British Journal of Surgery,* XXXI [1943], 38) writes: "In the same year, 1628, a Professor at Padua University, Johannes Colle, who may have known before of Harvey's work, did seriously suggest transfusion as a practical proposition, though he did not perform it." Dr. Keynes also discusses in detail the claim of Francesco Folli, a Florentine physician, to have been the pioneer, a claim which, he says, may be firmly rejected. This article, to which I shall refer again, is by all means the best history of the subject, and particularly illuminating in the section on the seventeenth century, since, as every student of Sir Thomas Browne knows, Keynes is expert in that field also.

It will probably never be determined who, in the post-Harveian period, first actually attempted blood transfusion. According to John Aubrey's *Lives*, not always the most reliable of sources, the honor belonged to Francis Potter, of whom we have heard in passing on two occasions. A divine, he was also a practical mechanic, inventor of various kinds of compasses and quadrants, as well as of one of the instruments for drawing in perspective, later reinvented by Wren. Thanks in part to Aubrey, who showed a number of his inventions to the Royal Society, Potter was elected a member not long after the chartering. Certainly if Aubrey's account is accurate, Potter's experiments preceded those of the Oxford group and of the Royal Society by several years, although not one of them was successful. Aubrey wrote:

Memorandum, that at the Epiphanie, 1649, when I was at his house, he then told me his notion of curing diseases, etc. by transfusion of bloud out of one man into another, and that the hint came into his head reflecting on Ovid's story of Medea and Jason, and that this was a matter of ten years before that time.[8]

This would date the first Potter experiments as early as 1639. Aubrey continues:

About a yeare after [presumably after their conversation, e.g., 1640] he and I went to trye the experiment, but 'twas on a hen, and the creature to little and our tooles not good: I then sent him a surgeon's lancet. Anno . . . I recieved a letter from him concerning this subject, which many yeares since I shewed, and was read and entred in the bookes of the

[8] *"Brief Lives," Chiefly of Contemporaries, Set Down by John Aubrey, between the Years 1669 & 1696,* ed. Andrew Clark (Oxford, 1898), II, 163–69. Potter appears in Pepys' *Diary* in a quite different connection. Like many others, Pepys was greatly interested in the only book Potter ever published, *An Interpretation of the Number 666* (Oxford, 1642), which was widely read in England in the year 1666, since it professed to show that "this number is an exquisite and perfect character, truly, exactly, and essentially describing the state of Government to which all other notes of Antichrist doe agree." On February 18, 1666, Pepys went to his bookseller's to order a copy which, he noted on November 4, 1666, "pleases me mightily."

Royall Societie, for Dr. Lower would have arrogated the invention to himself.[9]

Aubrey at this point quotes the letter, with Potter's statement that he had continued attempts at transfusion on hens, but without success, "for I cannot, although I have tried divers times, strike the veine so as to make him bleed in any considerable quantity. . . . I cannot procure above 2 or 3 drops of blood to come into the pipe or the bladder." One mystery to which neither Potter nor Aubrey offered any clue is why Potter limited his experiments to hens, those almost bloodless creatures.

Although Pepys naturally did not realize the interrelations, he had been present at various experiments and had heard reports of others, which we now recognize as integral parts of two series leading directly to the first successful transfusions. On January 22, 1666, the Royal Society held its first meeting after the plague. Pepys was not present, but joined a group at supper:

I back presently to the Crowne taverne behind the Exchange by appointment, and there met the first meeting of Gresham College since the plague. Dr. Goddard did fill us with talke, in defence of his and his fellow-physicians going out of towne in the plague-time; saying that their particular patients were most gone out of towne, and they left at liberty; and a great deal more &c. [Pepys may be pardoned his impatience with Goddard, since he himself had remained in London throughout most of the plague.] But what, among other fine discourse pleased me most, was Sir G. Ent about Respiration; that it is not to this day known, or concluded among physicians, nor is to be done either, how the action is managed by nature, or for what use it is. Here late till poor Dr. Merriot was drunk, and so all home, and I to bed.

[9] The letter is not dated, although it seems to have been enclosed with another that follows, dated December 7, 1652. This agrees with a statement in Birch, II, 361, that on April 29, 1669, "Mr. Aubrey produced a letter written him by Mr. Francis Potter in 1652, signifying, that at that time the writer of it had made some trials of the transfusion of blood." In his letter to Aubrey, Potter drew a "rude figure" of the ivory pipe and bladder which he used for transfusion.

Sir George Ent [10] did not exaggerate. At this time there was no agreement among physicians and scientists about the function or process of respiration. Not even Robert Boyle could answer such questions. But ever since they had developed the air pump, rightly one of the treasures and showpieces of the Royal Society, Boyle and Hooke had been experimenting upon problems of air. Marie Boas writes of this period of Boyle's career:

With this [the air pump] Boyle and his assistants performed a series of brilliant and important experiments, ingenious in conception and far-reaching in their scientific consequences, on the physical properties of air, especially its elasticity, proof of its weight and of the validity of the Torricellian experiment, experiments on dynamics *in vacuo,* on the transmission of light, sound, magnetism, and so on, and experiments on the possibility of combustion and respiration *in vacuo.*[11]

Two years earlier, on an occasion of which we shall hear, Pepys might have joined the aristocrats and the King in "laughing at Gresham College . . . for spending time only in weighing of ayre, and doing nothing else since they sat." He knew by this time that the Society had done many other important things and that the frequent reports on the nature of air, led by Hooke and Boyle, were of great importance. In many cases the experiments performed before the Royal Society were repetitions of those already made at Oxford by Boyle, Lower, Wren, and others of their associates. Some had been made privately by Hooke in London before they were demonstrated publicly at Society meetings. The tests were always under the aegis of one or the other of those two great innovators, who laid the basis for our modern knowledge of the nature of air.[12]

[10] Ent was an M.D. and president of the College of Physicians, 1670–75, 1682–84. He had only recently been knighted by the King in the Harveian Museum, after his delivery of Harveian lectures on anatomy.

[11] *Robert Boyle and Seventeenth-Century Chemistry* (Cambridge, 1958), p. 44.

[12] Some idea of the extensiveness of Boyle's studies of the kind of experiment reported by Pepys may be seen in his *Second Continuation of Physico-Mechanical Experiments touching the Spring and Weight of the Air (Works* [London, 1744], IV, 161–95, and Appendix, 196–205), in which he reported dozens of in-

Since this subject is peripheral to that of blood transfusion, even though empirically essential, I shall limit my discussion to one experiment Pepys witnessed and two others of which he certainly heard at meetings of the Society. On March 22, 1665, Pepys wrote:

Then to Gresham College, and there did see a kitling killed almost quite, but that we could not quite kill her, with such a way; the ayre out of a receiver, wherein she was put, and then the ayre being in upon her revives her immediately; nay, and this ayre is to be made by putting together a liquor and some body that ferments, the steam of that do do the work.

As so often, Pepys' account makes the experiment more confusing than it actually was. The problem at issue had to do with finding some means for divers to breathe under water. The process consisted in showing that a living creature, placed in an evacuated container, would smother, but that when air was introduced, it would be revived by artificial resuscitation.[13]

A famous experiment on artificial respiration was suggested to the Royal Society by Hooke, who had already performed it privately.[14] Birch records briefly on May 9, 1667, an experiment by Lower on the *ductus thoracicus* of a dog, "to open the thorax of a dog, and to keep him alive with blowing into the lungs with bellows." At the meeting on May 23 "Hooke

stances in which fruit, vegetables, birds, frogs, mice, flies, and the like were tested in his receiver.

[13] The experiment was described by Birch, II, 25. Apparently Pepys had arrived late at the meeting, since the experiment had been tried on a bird before the kitten was introduced into the evacuated receiver. This is not the same experiment (though it has similarities with it) as an earlier one referred to in the satirical "Ballad of Gresham College," which I shall mention in the third essay.

[14] Hooke's experiment was a landmark in the history of the study of respiration. *Garrison and Morton's Medical Bibliography* comments, no. 916, p. 81. "By blowing air from a bellows over the exposed lungs of a dog, Hooke proved that respiratory motion is not necessary to maintain life, but that the essential feature of respiration lies in certain blood changes in the lungs." Oldenburg's letter to Boyle is in Boyle, *Works*, V, 369. This experiment, as well as nearly all the others treated in this essay, is also described in R. T. Gunther, *The Life and Works of Robert Hooke*, in *Early Science in Oxford* (Oxford, 1923–45), Vol. VI, but since the accounts are largely drawn from either Birch or *Phil. Trans.* I have not cross-referenced them.

moved, that some experiments be made, to find whether it be the supply of fresh air, or the motion of the lungs, that keeps animals alive"—again the basic problem of the function of respiration and how it is performed. He, too, referred to Lower's experiment and was instructed to prepare a similar one for the Society. When I say that Pepys must have known about this experiment, even if he did not witness it, I speak advisedly, for I have noticed no other experiment so frequently discussed at meetings, from the time it was first mentioned on May 23 until it was successfully performed on October 10, 1667. In spite of reiterated demands the experiment was postponed from meeting to meeting, on one occasion because of the visit of the Duchess of Newcastle, on another because Dr. King, who was to assist Hooke, did not appear, on others because the equipment did not prove satisfactory—as was the case when it was attempted unsuccessfully on July 18. The successful experiment on October 10 is described in Birch, in the *Philosophical Transactions,* and elsewhere. I follow an account given by Oldenburg in a letter to Boyle, written on October 15, 1667:

On Thursday last we repeated at our meeting, that notable experiment of opening a dog, and laying bare his lungs, and blowing into him with bellows, keeping him thus alive as long as we pleased; which occasioned some discourse about the nature and use of respiration; some of the company declaring, that this experiment made out, what was not the use of respiration; others intimating, that it seemed to teach, that the principal end of respiration was the discharging of fumes of the blood; the animal keeping alive, and lying still, as long as the lungs were supplied, and kept extended with fresh air; and falling soon into convulsions when we ceased to blow.[15]

As might have been expected, the conversation on this occasion had been led by the vocal Sir George Ent. Birch noted: "Sir George Ent, reflecting on this experiment, said that it shewed what was not the use of respiration, but not what it

[15] Boyle, *Works,* V, 369.

was." [16] Beginning with the next meeting of the Society, this was followed by discussion of another experiment reported at Oxford by Lower who had already performed it with Boyle: "Dr. Lower offered to make at the next meeting the experiment of breaking the nerves of the diaphragm in a dog, to make him broken-winded, and the operator was ordered to provide a dog for that purpose." The "broken-winded dog" [17] came to vie in fame with the dog preserved by bellows. The "operator"—a euphemism for Robert Hooke—again was ordered to provide equipment, including presumably the dog. Our sympathies go out to that great scientist, man of all work, jack-of-all-trades, dogcatcher-extraordinary to the Royal Society of London.

Important though the experiments on respiration were in laying one basis for blood transfusion, they do not lie as immediately behind the first transfusions as did another group of experiments, three of which Pepys witnessed. These had to do with the infusion of opiates and poisons into the veins of animals. On May 16, 1664, Pepys had noted:

By invitation to Mr. Pierce's, the chyrurgeon . . . to see an experiment of killing a dogg by letting opium into his hind leg. He and Dr. Clerke did fail mightily in hitting the vein, and in effect did not do the business after many trials; but with the little they got in, the dogg did presently fall asleep, and so lay until we cut him up, and a little dogg also, which they put it down his throate; he also staggered first, and then fell asleep and so continued. Whether he recovered or no, after I was gone, I know not.[18]

On March 15, 1665, Pepys again reported an attempted infusion: "Anon to Gresham College, where, among other good discourse, there was tried the great poyson of Maccassa [the upas tree] upon a dogg, but it had no effect all the time we sat

---

[16] II, 188–89. The experiment is mentioned by Birch, II, 173, 176, 181, 188, 189, 197, 200. In addition there is an account in *Phil. Trans.*, II (October 21, 1667), 539, and II (November 11, 1667), 544.

[17] These experiments are mentioned by Birch, II, 200, 201, 207.

[18] This experiment does not seem to have been at the Royal Society, since it is not mentioned in Birch or any other source I have examined.

there." [19] The chief excitement of that meeting proved to be
the discovery that Dr. Charleton, to whom the poison had been
entrusted, had taken it home with him—in opposition to the
regulations of the Society. He was sternly summoned to appear
at the next meeting, presumably to be disciplined, though his
real discipline was to come in the form of a satire of which we
shall hear.

The third infusion was tried on April 19, 1665, at which
time Pepys wrote:

To Gresham College, where we saw some experiments upon a
hen, a dogg, and a cat, of the Florence poyson. The first it
made for a time drunk, but it come to itself again quickly; the
second it made vomitt mightily, but no other hurt. The third I
did not stay to see the effect of it, being taken out by Povy.

If Pepys had remained, he would have learned, as the minutes
show, that the infusion had little effect on the kitten save that
it became drowsy and "slabbered" at the mouth. Pepys was
undoubtedly aware, as were other members, that the Florentine poison had been presented to the Society by the King,
through Sir Robert Moray, with His Majesty's hope that some
experiments would be made by the virtuosi. Perhaps Pepys'
mind was occupied at that meeting with the fact that he himself had made his first report (which he did not mention in the
Diary) about certain problems in connection with pendulum
watches as they were used at sea.

Such experiments of infusing drugs or poison into the veins
of animals had been developed at Oxford by Christopher
Wren, sometimes joined by Boyle himself, who wrote of his
part in one of them:

I may here mention some late experiments to show the effects
of liquid poisons conveyed immediately into the blood; and

[19] In the minutes for the preceding meeting, Birch (II, 21) notes: "Mr. Graunt
produced a box of Macassar poison, which was ordered to be tried at the next
meeting, by dipping a needle in the poison, and pricking some dog, or cat, or
pullet with it." On March 15 (II, 23) the notation is: "The experiment of trying to poison a dog with some of the Macassar powder, in which a needle had
been dipped, was made, but without success."

particularly that famous one of Mr. Christopher Wren, who contrived a new way of injecting them. I procured a large dog, into the vein of whose hinder leg we conveyed, by a syringe, a small dose of a warm solution of opium in sack. The effects whereof became manifest as soon as we cou'd loose the dog from the cords wherewith his feet were tied; for he immediately began to nod and reel as he walk'd; whereupon, to preserve his life, I order'd him to be kept awake by whipping. . . . The same gentleman, at another time, injected in the same manner about two ounces of *Vinum benedictum,* which operated so violently that it soon killed the dog.[20]

The syringe that Wren developed was the ancestor of the modern hypodermic needle. There is plenty of evidence that these experiments by Wren were recognized both at Oxford and in London as the immediate predecessors of blood transfusion. In Number 28 of the *Philosophical Transactions* for October 21, 1667, Oldenburg, resenting a French claim for priority in transfusion, called attention to Number 7 of the *Transactions* for December, 1665, in which he had acquainted the world "how many years since Dr. Christopher Wren proposed the experiment of Infusion into Veins. And this was hint enough for the R. Society, some while after to advance Infusion to Transfusion." A similar claim for Wren's experiments as the immediate predecessors of those on transfusion was made by Thomas Sprat, in *The History of the Royal Society of London:*

He [Wren] was the first Author of the Noble Anatomical Experiment of Injecting Liquors into the Veins of Animals. An Experiment now vulgarly known; but long since exhibited to the Meetings at Oxford, and thence carried by some Germans, and published abroad. By this Operation divers Creatures were immediately purg'd, vomited, intoxicated, kill'd or reviv'd, according to the quality of the Liquor injected: Hence

[20] *The Usefulness of Philosophy* in *Philosophical Works,* ed. Peter Shaw (London, 1738), I, 38–40. In notes to this passage and the one following, Boyle mentions a number of other experiments, particularly a group made by Dr. Friend "with a view to shew the manifest alterations that different medicines would cause in the blood." He lists some thirteen different medicines which had been tested by infusion into the blood stream.

arose many new Experiments, and chiefly that of transfusing Blood, which the Society has prosecuted in sundry instances, that will probably end in extraordinary Success.[21]

II

The simplest and most succinct account of the first transfusion recorded by the Royal Society, mentioned by Pepys and reported in the introduction to this essay, is that given under date of November 14, 1666, in the minutes preserved in Birch's *History:*

The experiment of transfusing the blood of one dog into another was made before the Society by Mr. King and Mr. Thomas Coxe, upon a little mastiff and a spaniel with very good success, the former bleeding to death, and the latter receiving the blood of the other, and emitting so much of his own, as to make him capable of receiving the other. It was ordered that the whole method, and all the particulars of the operation, should be fully described, and brought in at the next meeting.

Although "Mr. King and Mr. Thomas Coxe," both of whom were doctors, played important parts in the history of the Royal Society, on this occasion of the first blood transfusion ever performed in public, they were not innovators, but rather glorified laboratory assistants, performing an experiment they had not devised, in which they had been coached by Robert Boyle and the man responsible for this important innovation in the history of medicine, Dr. Richard Lower, who had performed the same operation earlier at Oxford. Lower was one of the most important English physiologists, in later years probably the most noted physician in Westminster and London. At Oxford he had been closely associated with John Wallis, whom he assisted in anatomical research, and was on terms of friendship with Boyle. His classic work, *Tractatus de Corde,*[22] a study

---

[21] (London, 1667), p. 317.

[22] *Tractatus de Corde item De Motu & Colore Sanguinis et Chyli in eum Transitu* (London, 1669). M. W. Hollingworth published an English translation

of the heart and of the circulation of the blood, ranks second in importance in England to Harvey's great contribution. In this work, published in 1669, Lower devoted a chapter to the problem of transfusion of the blood, including an account of the operation between two dogs he had performed at Oxford in February, 1665—the first known transfusion in the history of medicine. There is no question that he was consciously following Wren's experiments in infusion.

The instructions Boyle sent to the Royal Society constitute an English version of sections later published as Chapter IV of *De Corde*, which consists in part of letters between Lower and Boyle, together with Lower's defense of his own priority over the French in the operation of blood transfusion. Here are Lower's instructions for "The Method in Transfusing the Bloud Out of One Animal into Another. . . . First Practised by Doctor Lower in Oxford, and by Him Communicated to the Honourable Robert Boyle, Who Imparted It to the Royal Society." I give the first part of this historical document in the words in which it was communicated to the Royal Society and paraphrase the rest:

First. Take up the Carotidal Artery of the Dog or other Animal, whose Bloud is to be transfused into another of the same or a different kind, and separate it from the Nerve of the Eight pair, and lay it bare above an inch. Then make a strong Ligature on the upper part of the Artery, not to be untied again; but an inch below, *videl.* towards the Heart, make another

of the section on transfusion, "Blood Transfusion by R. Lower in 1665," in *Annals of Medical History*, X (1928), 213-20. A facsimile edition of the original, together with a complete translation and comment by K. J. Franklin, appeared in Gunther, *Early Science*, Vol. IX. The translation from which I quote in the text (whether Lower's, Boyle's, or Oldenburg's) appeared in *Phil. Trans.*, I (December 17, 1666), 353-57.

In Chapter IV of *De Corde* (ed. Franklin, not paginated) Lower wrote: "For many years at Oxford I saw others at work, and myself, for the sake of experiment, injected into the veins of living animals various opiates and emetic solutions, and many medicinal fluids of that sort. . . . I finally showed this new experiment [of transfusion] at Oxford towards the end of February 1665, in an interesting demonstration and under the most happy cicumstances. There were present the learned Doctor John Wallis, Savilian Professor of Mathematics, Thomas Millington, Doctor of Medicine, and other Doctors of the same University,"

Ligature of a running knot, which may be loosen'd or fastned
as there shall be occasion. Having made these two knots, draw
two threds under the Artery, and put in a Quill, and tie the
Artery upon the Quill very fast by those two threds, and stop
the Quill with a stick. After this, make bare the Jugular Vein
in the other Dog about an inch and a half long; and at each
end make a Ligature with a running knot, and in the space
betwixt the two running knots drawn under the Vein two
threds, as in the other; then make an Incision in the Vein, and
put into it two Quills, one in the descendent part of the Vein,
to receive the bloud from the other Dog, and carry it to the
Heart; and the other Quill put into the other part of the Jugular Vein, which comes from the Head (out of which, the second Dogs own Blood must run into Dishes). These two Quills
being put in and tyed fast, stop them with a stick till there be
occasion to open them.

We may visualize the two dogs, tied on their sides, facing
one another. Since they cannot be brought near enough that
two quills will suffice, the operator is instructed to add others
(probably two or three), depending upon the sizes of
the dogs, to form an unbroken conveyance from one animal to
the other. This done, the operator slips the running knot, and
the arterial blood gushes forth, passing through the tubes
into the vein of the other dog. One other piece of instruction is
given by Lower, which I mention because I shall return to it in
a different connection: the operator must be careful to observe
the pulse beyond the quill in the dogs' jugular veins, to make
sure that it is beating regularly.

Such was Lower's account of the experiment which, he said,
"I have tried several times before several in the Universities."
And such was the method followed when the experiment was
successfully repeated before the Royal Society by Drs. King
and Coxe. Although they were not the originators, they deserved the applause they received, in part because of their deft
technique—very different from the bungling Pepys had witnessed by a technician unable to find the vein in a dog—and
also because they were to continue transfusion experiments
with important results. In the original experiments at Oxford

and in London, it will be noticed, the emittent dog was deliberately permitted to die by the removal of most of his blood. In later operations, several of which King and Coxe performed together, the emittent animal was usually kept alive, only a small portion of his blood being removed. In addition, after experiments with using the crural rather than the carotid artery, King reported to the Royal Society "An Easier and Safer Way of Transfusing Blood Out of One Animal into Another, viz., by the Veins, without Opening Any Artery of Either." We shall notice other evidences of King's and Coxe's experiments presently, but I turn momentarily to Robert Boyle, who stimulated these and other physicians and scientists to many of their discoveries.

Boyle's interest in problems relating to blood may be seen in various of his works, such as *Memoirs for the Natural History of Human Blood*.[23] The extensiveness of his inquiry into many aspects of the subject shows in the long, leisurely indexes to early editions of his works. In the Peter Shaw edition, for example, entries on "Blood" run to over two columns. I shall pause over only one brief paper which seems to have done more to stimulate widespread interest in problems involved with blood transfusion than any other one publication: this was a highly suggestive list of "queries" about animal transfusion. When these were published in the *Philosophical Transactions* for February 11, 1666,[24] they were headed: "Tryals Proposed by Mr. Boyle to Dr. Lower, to Be Made by Him for the Improvement of Transfusing Blood Out of One Live Animal into Another." Boyle indicates that he is publishing the queries at the instigation of "the worthy Doctor, to whom they were addressed, who thinks they may excite and assist others in a matter, which, to be well prosecuted, will require many hands." And so indeed they did. The basic question behind the queries was "whether by transfusion of blood,

---

[23] In *Works* (1744), IV, 161–95, and Appendix, 196–205.

[24] I, 385–88. Boyle indicates here that these were notes he made much earlier. The queries were also published, without the initial statements, in *The Usefulness of Experimental Philosophy* (*Phil. Works*, I, 39–41).

the disposition of individual animals of the same kind, may not be much altered." Boyle was raising two main problems: the extent to which transfusion may affect native characteristics and whether or not it will have any effect upon acquired characteristics. Among his many questions are: Whether a fierce dog, by being quite new stocked with the blood of a cowardly dog, may not become more tame or vice versa? Whether a transfused dog will recognize his master? Whether characteristics peculiar to a breed (e.g., the scent of bloodhounds) will be abolished or impaired if a spaniel's blood is transfused into a bloodhound? Whether rejuvenation will occur if an old, feeble dog is given the blood of a young, vigorous one? Whether the blood of one animal may be safely transferred into another, as a calf into a dog, or a cold animal, as a fish, frog, or tortoise, into a hot animal, and vice versa? The list concludes with a general inquiry: "Whether, by frequent transfusion, something tending to a degree of change of species may be accomplished?" Boyle raised the questions. He did not pretend to answer them. Yet as one reads between the lines, it seems that Boyle felt that, important though transfusion might prove to be, its effects would be physiological, rather than what we would call psychological. Particularly in passages in which he makes passing comparisons with grafting of trees and plants, he seems to think that transfusion will not ultimately result in a crossing of species.

The queries provoked great interest at home and abroad, and various attempts were made to answer at least some of them. On May 6, 1667, Dr. King reported a successful cross transfusion, during which he had introduced into a calf forty-nine ounces of blood from a sheep.[25] In the same number of the *Transactions* Dr. Coxe reported "An Account of Another Ex-

---

[25] II, 449–51. According to a notice in the same number of *Phil. Trans.* cross transfusion had been successfully performed in France a month or so earlier. Dr. Denis, of whom we shall hear more, reported in a letter dated April 2, 1667, that he and his colleagues had transfused the blood of three calves into three dogs. Coxe's account of the sound and mangy dogs appears on pp. 451–52 of the same number.

periment of Transfusion, viz. of Bleeding a Mangy into a Sound Dog." His subjects had been "an old Mungrell Cur, all overrun with the Mainge," and "a young Land-Spaniel, of about the same bigness." He had transfused from fourteen to sixteen ounces of the blood of the infected dog into the sound one, and reported that no alteration was observed in the sound animal, but that the mangy one "was in about 10 dayes or a fortnight's space perfectly cured." [26]

In 1667 and 1668 so many transfusions between animals, of the same or different kinds, took place that to collect them would result only in a dull catalogue of common things. More than a dozen, made in England or reported from the continent, are mentioned in Birch's *History* or in the *Philosophical Transactions*. I shall include only one reply to Boyle's queries, which Oldenburg picked up from the *Giornale de litterati*, reporting, as of May 8, 1667, the transfusion of a lamb's blood into a spaniel, thirteen years old, which had been deaf for three years and was so feeble that it walked only with difficulty. [27] Within an hour after the transfusion the animal leapt from the table fairly briskly. Two days later it was running the streets and eating greedily. Within two weeks it had been cured of deafness. Such at least was the report to the Royal Society. "They say miracles are past," said Lafeu in *All's Well That Ends Well*. Shakespeare lived just too early to know about blood transfusion.

[26] If Coxe seems to be confusing cart and horse, I may point out that Boyle had phrased his query in the reverse order from what we might expect, asking about the effect of the blood of an infected dog upon a sound animal. Coxe made no pretense of attributing the cure of the mangy dog to blood transfusion. He considered it a result of "the considerable evacuation [the mangy dog] made," which seems to have been a usual prescription for mange.

[27] *Phil. Trans.*, III (December 14, 1668), 840–41. An analogous case had been reported earlier from France, without so much corroborative detail. *Phil. Trans.*, II (June 3, 1667), 479: "M. Gayant shew'd the Transfusion of the Bloud, putting that of a Young Dog into the Veins of an Old, who, two hours after, did leap and frisk; whereas he was almost blind with age, and could hardly stirr before."

III

Pepys will again be our guide as we pass from earlier experiments on transfusion between animals to the first English blood transfusion into a human being. On November 21, 1667, he wrote:

Did much business till after candlelight, and then my eyes beginning to fail me, I out and took coach to Arundell House, where the meeting of Gresham College was broke up; but there meeting Creed, I with him to the taverne in St. Clement's Churchyard. . . . We fell to other discourse, and very good; among the rest they discourse of a man that is a little frantic, that hath been a kind of minister, Dr. Wilkins saying that he hath read for him in his church, that is poor and a debauched man, that the College have hired for 20s. to have some of the blood of a sheep let into his body, and it is to be done on Saturday next. They purpose to let in about twelve ounces; which, they compute, is what will be let in in a minute's time by a watch. They differ in the opinion they have of the effects of it; some think it may have a good effect upon him as a frantic man by cooling his blood, others that it will not have any effect at all. But the man is a healthy man, and by this means will be able to give an account what alteration, if any, he do find in himself, and so may be useful.

The persistent question of possible physiological and psychological effects of transfusion led Dr. Whistler to tell "a pretty story" of "Dr. Caius, that built Keys College; that, being very old, and living only at that time upon woman's milk, while he fed upon the milk of an angry, fretful woman, was so himself; and then, being advised to take it of a good-natured, patient woman, he did become so, beyond the common temper of his age. Thus much nutriment, they observed, might do."

Since, because of the pressure of his duties, Pepys had not been able to attend Society meetings as regularly as he had done at first, he may not have been aware of earlier attempts to find a subject for human transfusion. On October 17 "Mr. Oldenburg moved, that the experiment of transfusion of blood might be prosecuted and considered, in order to try it with safety upon men, it having been already practised at Paris." At the next meeting on October 24 Sir George Ent suggested that they procure as patient "some mad person in the hospital of Bethlem." A distinguished committee was appointed to approach the Keeper of Bedlam, Dr. Thomas Allen, who, they reported, "scrupled" to permit the trial upon any of the mad patients. Ent was then requested to call a meeting of physicians at his home in order "to consider together, how this experiment might be most conveniently and safely tried." [28] Possibly the suggestion about Arthur Coga,[29] the patient who received the transfusion, was made at that time. Pepys has mentioned the fact that Coga was personally known, at least to John Wilkins. Pepys was unable to witness the transfusion on Saturday, November 23, 1667. As comments in the *Diary* indicate, this was a tense period in his professional life, when he had reason to expect his dismissal from office. Lord Brouncker, President of the Royal Society, was also regretfully absent from the transfusion, involved as he was in the same crisis. Pepys never paid the Society higher tribute than when he wrote, on the evening he mentioned the coming transfusion, "Their discourse was very fine; and if I should be put out of my office, I do take great content in the liberty I shall be at, of frequenting these gentlemen's company."

There are various accounts of the first English transfusion into a human being, including a brief one by Richard Lower, who performed the operation.[30] I shall follow a letter from

[28] Birch, II, 201–2. Ent's proposal "being seconded by divers other physicians of the Society, Dr. Lower, Dr. King, Mr. Thomas Coxe, and Mr. Hooke were desired to speak with Dr. Allen, physician to Bethlem, about the execution of this trial, and to let him know the opinion declared in the society concerning it" (*ibid.*, p. 204): On October 31, when Allen's negative reply was reported, "it was ordered, that he should be desired by Mr. Hooke to give a meeting at Sir George Ent's house on the Monday following to some of the physicians of the Society . . . to consider together, how this experiment might be most conveniently and safely tried."

[29] According to Keynes, p. 43, Coga, age about thirty-two, was an indigent Bachelor of Divinity of Cambridge and brother to the master of Pembroke.

[30] Lower's is given in Chapter IV of *De Corde* (ed. Franklin, not paginated). A report appeared in Birch, II, 214–16. The one in *Phil. Trans.*, II (December

Oldenburg to Boyle because it includes all the details in the other accounts and gives some interesting additional touches:

On Thursday next, God willing, a report will be made of the good success of the first trial of transfusion practised on a man, which was by order of the Society, and the approbation of a number of physicians, performed on Saturday last in Arundel House, in the presence of many spectators, among whom were Mr. Howard and both his sons, the bishop of Salisbury, four or five physicians, some parliament men, etc., by the management and operation of Dr. Lower and Dr. King, the latter of whom performed the chief part with great dexterity, and with so much ease to the patient, that he made not the least complaint, nor so much as any grimace during the whole time of the operation; in which the blood of a young sheep, to the quantity of about eight or nine ounces by conjecture, was transmitted into the great vein of the right arm, after the man had let out some six or seven ounces of his own blood. All which was done by the method of Dr. King's, which I published in Num. 20 of the Transactions, without any change at all of it, save only in the shape of one of the silver pipes, for more conveniency. Having let out, before the transfusion, into a porringer, so much of the sheep's blood, as would run out in about a minute (which amounted to twelve ounces) to direct us as to the quantity to be transfused into the man, he, when he saw that florid arterial blood in the porringer, was so well pleased with it, that he took some of it upon a knife, and tasted it, and finding it of a good relish, he went the more couragiously to its transmission into his veins, taking a cup or two of sack before, and a glass of wormwood wine and a pipe of tobacco after the operation, which no more disordered him, both by his own confession, and by appearance to all bystanders, than it did any of those that were in the room with him. The pipe being taken out of the man, the blood of the sheep ran a very free stream, to assure the spectators of an uninterrupted course of blood.

So far, Oldenburg's account agrees in every way with the paper Dr. King contributed to the *Philosophical Transactions* and

---

9, 1667), 557–59, was communicated by Dr. King, who assisted Lower. King sent another account in a letter to Boyle, dated November 25, 1667 (see *Works* [1744], V, 638). Oldenburg's letter (*ibid.*, V, 371–73) was written on November 25, 1667.

the letter he wrote to Boyle. In the section that follows are details that appear in none of the other accounts:

The patient found himself very well upon it, his pulse better than before, and so his appetite. His sleep good, his body as soluble as usual, it being observed, that the same day of his operation he had three or four stools, as he used to have before. This morning our president (who by very pressing business could not be present in Arundel House) and I went to see him pretty early, and found him a bed, very well, as he assured us, and more composed, as his host affirmed, than he had been before, he being looked upon as a very freakish and extravagant man; who hath studied at Cambridge, and is said to be a batchelor of divinity, called Arthur Coga, an indigent person, and receiving a guinea for undergoing the experiment; which reward maketh him willing to have it repeated upon him, wherein he will easily be complied with, and that, I think, before the end of this very week, if circumstances shall persuade it.

Pepys had not witnessed the operation but he was to have an opportunity of seeing the patient. On November 30 he attended a meeting of the Royal Society for the election of officers and was flattered that he "was near being chosen of the Council . . . above all things, I could wish it; and do take it as a mighty respect to have been named there." After the meeting, as usual, a group of members dined together. Pepys wrote:

I choosing to sit next Dr. Wilkins, Sir George Ent, and others whom I value, there talked of several things. . . . Here, above all, I was pleased to see the person who had his blood taken out. He speaks well, and did this day give the Society a relation of it in Latin, saying that he finds himself much better since, and as a new man, but he is cracked a little in his head, though he speaks very reasonably and very well. He had but 20s. for his suffering it, and is to have the same again tried upon him: the first sound man that ever had it tried on him in England, and but one that we hear of in France, which was a porter hired by the virtuosos.

Two details are added to this by King in his letter to Boyle. The physician felt that Coga exaggerated the improvement he felt in his own health, "enlarging more upon the benefits, he thinks, he hath received of it, than we think fit to own as yet." The other addition affords one of the most amusing details of the whole episode. In a postscript King told Boyle that during the period of conversation, Coga was asked "why he had not been given some other creature's blood" rather than that of a lamb. Coga replied: "Sanguis ovis symbolicam quamdam facultatem habet cum sanguine Christi, quia Christus est agnus Dei." A symbolic relationship indeed the quondam minister read into the blood of *his* lamb and that of the Lamb of God. If Arthur Coga, "cracked a little in his head," continued to preach, his favorite passages and hymns for lower-class revivals may well have been those describing sinners who had been "washed in the Blood of the Lamb."

The second operation upon Arthur Coga—and the last that was to occur in England until the nineteenth century—was performed two and a half weeks later, on December 12, 1667, under less propitious auspices than the first because of the widespread interest caused by that. The minutes of the Society, as reported in Birch, read:

The second experiment of transfusion was made by Dr. King upon Mr. Arthur Coga, by taking from him eight ounces of blood, and transmitting into him, by guess, about fourteen ounces of sheep's blood. Dr. King was desired to bring in an account of it to be registered. This experiment being made in a great crowd of spectators, which would not admit of that exactness, which was designed, the physicians of the society were requested to take an opportunity of making this experiment more exactly by weighing the emittent animal before and after transfusion.

As on the preceding occasion, Coga appeared in person before the Society a week later on December 19:

Mr. Coga being introduced gave an account of the effects of the experiment of transfusion repeated upon him, viz. that he found himself very well at present, though he had been at first somewhat feverish upon it; which was imputed to his excess in drinking too much wine soon after the operation.[31]

The instructions to the physicians at the meeting of December 12 would indicate that another transfusion was intended, as does a statement made by Lower in *De Corde* about

the case of a certain A.C., who was the subject of a harmless form of insanity. I superintended the introduction into his arm at various times of some ounces of sheep's blood at a meeting of the Royal Society, and that without any inconvenience to him. In order to make further experiments on him with some profit also to himself, I had decided to repeat the treatment several times in an effort to improve his mental condition; he, on the other hand, consulted his instinct rather than the interests of his health, and completely eluded our expectations.[32]

Arthur Coga may or may not have written a letter attributed to him by Henry Stubbe, who was constantly on the warpath so far as the Royal Society and the New Science in general were concerned. In his animadversions upon Joseph Glanvill's *Plus Ultra,* a defense of the Society, Stubbe published a letter supposedly written by Coga, implying some vaguely unhappy results of the transfusion, yet saying that he "hath more blood still at your service." [33] While it is possible that Coga wrote that letter of his own volition, I prefer to leave it for the present and return to it in my essay on satire

[31] The first account is in Birch, II, 225, the second, II, 227.

[32] The translation of *De Corde,* as I have said, is unpaginated. Lower's statement is about a page before the end of Chapter IV. His phrase, "at various times," is a little curious, since, so far as I can determine, he operated only twice. No record of the second—or other—operation appears in *Phil. Trans.* I shall return to the case of Arthur Coga in the last essay.

[33] See Stubbe, *A Specimen of Some Animadversions upon a Book, Entitled, "Plus Ultra"* . . . *Written by Mr. Joseph Glanvill, a Member of the Royal Society* (London, 1670), p. 179. This work is frequently bound with *The "Plus Ultra" Reduced to a Non Plus; or, A Specimen of Some Animadversions upon the "Plus Ultra" of Mr. Glanvill, wherein Sundry Errors of Some Virtuosi Are Disclosed* (London, 1670). The letter is reprinted by Franklin in his edition of Lower's *De Corde,* pp. xxiii–xxiv. I am greatly indebted to Mrs. Brand Blanshard for finding the Stubbe references for me in the Yale University Library.

directed against the Royal Society, for, whoever wrote it, I am convinced that satire lay behind it and that the reason Coga did not return for the other operations Lower had intended was that he had been taken in hand by "the Wits." But enough of Coga for the moment; we shall hear of him again on at least two occasions.

Interest in the possibility of human transfusion continued among members of the Royal Society for some time. On February 17, 1668, some two months after the Coga operation, a proposal was made to try transfusion upon "a poor distracted woman," who was really insane. Notes in Birch suggest that the plan was abandoned on account of the financial responsibility that might have been involved for the Society:

Dr. Clarke mentioned, that there was a poor distracted woman, who seemed to him a fit subject to try the transfusion upon; but that she not being provided for, it was to be feared, that she would lie upon the society's hands, after the experiment should be made upon her. He was therefore desired to speak with some of the officers of the parish, where she was then maintained, that in case they would continue to provide for her, the transfusion should be made upon her, as a means, which the physicians thought not unlikely to cure her.[34]

Since nothing further was said on the subject, we may assume that the officers of her parish refused to guarantee continuance of the dole for the "poor distracted woman."

The subject of transfusion continued to be raised occasionally at Society meetings until the middle of 1669. In entries beginning on March 18, 1669, we find the last specific proposal for a transfusion experiment, this time not on a human being:

Mr. Croune proposed an experiment, to try, whether an animal would be fed by blood alone transfused into it, viz. by en-

[34] Entries in Birch, II, 362, 377, follow up this proposal. On May 6, 1669, in answer to a question, Croone said he had done something but was not yet ready; on June 3 he complained that he "had wanted hitherto the hands of the operator, who was therefore ordered anew to defer no longer the furnishing of all the necessaries for these experiments, nor his attendance at such times, as should be convenient to Dr. Croune." On these occasions the "operator," I think, was not poor Robert Hooke.

closing two dogs in a box, and making the blood circulate from the one to the other by way of transfusion, feeding the one and not the other. He was desired to make the experiment, and Dr. Allen and Mr. Hooke to assist him in it.

What Dr. Croone called "circular transfusion" was mentioned by him or others at the meetings of May 6 and June 3, 1666, but the experiment was not performed before the Society. An earlier note in Birch's *History* for January 21, 1669, indicates the waning interest in the subject in England: "Notice being taken, that in foreign parts the experiments of injection and transfusion were much practised and improved, whereas they were neglected in England, where they were first invented." For an explanation of the change in temper of the once-venturesome pioneers—and also a suggestion for their reasons for wishing to experiment upon the insane—we must turn to France.

IV

I have deliberately violated chronology by discussing the first English transfusion upon a human being before dealing with earlier ones that had been performed in France. I have done so for various reasons: in part to carry on the English story from the first transfusions between animals to the English experiments upon a man; in part because, while at least three human transfusions had been performed in France earlier, the most famous—and infamous—one occurred later. Chiefly, I confess, I have changed the order for purposes of climax. As an avid reader of detective stories, I feel, with Dickens' Mr. Waterhouse, "Other things are all very well in their way, but give me Blood!" and echo the sentiments of a former maid of mine who used to say, "I love mellerdrammer, and the mellerer it is, the better!" If ever in the history of medical science there was a "meller" detective story, *l'affaire Saint Amant* will prove to be it. Let me at once pay delayed obeisance to France and make entirely clear that the first transfusion ever made on a human being occurred in Paris, some four months before the English experiment on Arthur Coga.[35]

The French pioneer was Jean Denis, doctor of the Faculty of Medicine at Montpellier and also a teacher of various branches of science and one of several physicians attached to Louis XIV. With him was associated Paul Emerez, a surgeon and teacher of anatomy and surgery. In one of his several communications to the Royal Society, Denis indicated that his starting point, so far as transfusion was concerned, had been the English experiments upon animals. His first, and one of his longest communications, was in a number of the *Philosophical Transactions*,[36] unusual in format, which appeared while Henry Oldenburg, Secretary of the Royal Society, was confined in the Tower, a circumstance about which Pepys wrote on June 25, 1667:

I was told yesterday, that Mr. Oldenburg, our Secretary at Gresham College, is put into the Tower, for writing newes to a virtuoso in France, with whom he constantly corresponds in philosophical matters; which makes it very unsafe at this time to write, or almost do any thing.

At the opening of his letter Denis courteously indicates his debt to English virtuosi: "You are justly intitled to a greater share than any other, considering that it was first spoken of in your Academy, & that the Publick is beholding to you for this

---

[35] The French cases have been studied in careful detail by Harcourt Brown and reported in "Jean Denis and Transfusion of Blood, Paris, 1667–1668," *Isis*, XXXIX (1948), 15–28. In this excellent article Professor Brown introduces evidence from many sources, including some hitherto unpublished manuscripts. I limit myself to letters of Jean Denis which were published in translation in the *Phil. Trans.*, II (July 22, 1667), 489–504—the version widely known among English readers—and some collaterial material from the *Transactions*.

[36] II (July 22, 1667), 489–504. As a result of Oldenburg's absence, the letter is extraordinarily garbled in pagination and continuity. Keynes says (p. 42): "Oldenburg was released just in time to have the letter supressed so that it is not included in most existing sets of the *Transactions*. It is now of great rarity." He adds in a footnote: "Copies are to be found in the library of the Royal Society, in Dr. John Fulton's library at Yale, and in my own." What seems the complete letter, though meandering through the number like Alph the sacred river, is in the photographic copy of Volume II of *Phil. Trans.* (Nieuwkoop-Amsterdam, 1963–64).

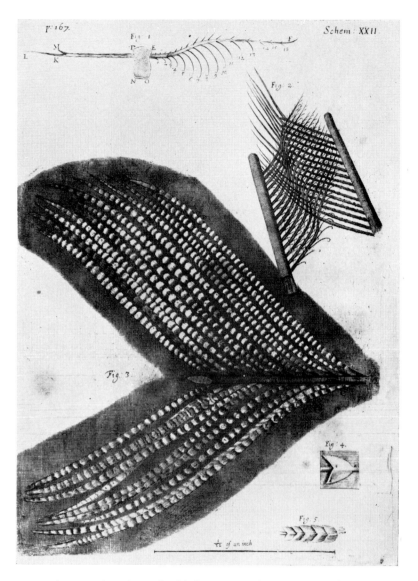

6. Feathers on the wings of a bird. (From Robert Hooke, *Micrographia*, 1667.)

7. Robert Boyle. (From a portrait by F. Kerseboom; courtesy of the Royal Society.)

as well as for many other discoveries." He mentions his own first experiments on animal transfusion, tried upon dogs about four months earlier. Sometimes he had transfused from the carotid or crural artery into the jugular vein, sometimes, as in the later English practise, from vein into vein. He had used "dogs both weak and strong, great and small." Among the nineteen on which he had experimented, not one had died. Most seemed to benefit from the experience. Following experiments with dogs alone, he had cross-transfused calf and dog with equal success. "We became confirm'd," he wrote, "in the Opinion that there was more ground to hope effects rather advantageous than hurtful to mankind, from this discovery of Transfusion of Blood." [37]

It seems clear that, before this time, Denis had determined that he would attempt human transfusion. He was urged to try it upon a condemned criminal, but decided against that and waited some time until he found a patient whose condition seemed to warrant it. "On the 15 of the Moneth," he noted,

we hapned upon a Youth aged between 15 and 16 years, who had for above two moneths bin tormented with a contumacious and violent fever, which obliged his Physitians to bleed him 20 times, in order to asswage the excessive heat. . . . Since the violence of this fever, his wit seem'd wholly sunk, his memory perfectly lost, and his body so heavy and drowsie that he was not fit for any thing.

An attempt at transfusion seemed warranted, and Denis, with the assistance of Emerez, proceeded to make it. Three ounces

[37] At this point, pp. 491–500, occurs a long and very interesting discussion of some of the objections which had been raised—presumably by other physicians and scientists—to blood transfusions in general. The three basic problems seem to have been: "The diversity of Complexions (which is founded in the Blood) supposeth so great a diversity in the several Blouds of different Animals . . . 2. Because Blood entravasated or removed out of its natural place, must necessarily corrupt . . . 3. Because the Blood issuing out of its proper Vessels, and being to pass through inanimated conveyances, such as are the Pipes or Tubes . . . must infallibly coagulate, and so coagulated descending to the heart must cause there a Palpitation, whereof death will be the speedy consequent." These points are discussed in detail, theoretically and practically. Denis insists that not one of these difficulties was encountered during his experiments.

of dark thick blood were drawn from a vein at the elbow and about eight ounces of arterial blood transfused from the carotid artery of a lamb. The effect was benign: the boy recovered from his lethargy, his appetite returned, he resumed his normal liveliness. The first human blood transfusion seemed abundantly justified. The second Denis acknowledged to have been "more by curiosity than by necessity." This was upon an unusually strong, healthy man of forty-five, whose occupation was carrying sedan chairs, which explains Pepys' comment about "a porter hired by the virtuosoes," though Pepys was wrong in thinking this the only human transfusion before Coga's. In this case, ten ounces were removed from the man's arm and twenty transfused from the crural artery of a lamb. Far from showing any ill effects from the experience, the patient—if such a hearty, amusing, and amused collaborator merited the term—sprang from the table, expertly cut the throat of the lamb and fleeced it, proceeded to drink up a good part of his fee, and was back at his usual strenuous work by two in the afternoon, insisting that he felt "more hearty" than before, though, as the doubting Denis noted, "whether by the new blood he had received six hours before, or by the quantity of wine he drank," was a moot question.

With the third French transfusion [38] we hear the first warning of a future when human blood transfusion was to cease for nearly a century and a half. The unfortunate result in this case seems to have been no fault of Denis and his associate. "Baron Bond, Son to the first Minister of State to the King of Sweden," as he was described in one of the fullest extant accounts,[39] was close to death, "afflicted with the complicated

---

[38] I consider here only the four cases Denis or another described in some detail. There may have been several others and there was certainly at least one, since in the account Denis sent the Royal Society about the lawsuit that followed the death of Saint Amant he spoke of a patient who was urging him to treat her by transfusion, "I have now before me a Paralytick Woman (a neighbour and friend to her, that was cured of the Palsey this way)" (*Phil. Trans.,* III [May 15, 1668], 715).

[39] *Phil. Trans.,* II (December 9, 1667), 557–64. This is based in large part upon a letter from Claude Gadroy to Abbé Bourdelot. There is another account in *Phil. Trans.,* II (October 21, 1667), 517–25.

distemper of an Hepatick Flux, a Lientery, and a bilious Diarrhea, accompanied with a very violent Feaver." He had been attended by at least four physicians who had "blouded, purged, and clystered him." As a last resort the somewhat reluctant doctors consented to a transfusion requested by relatives who had heard of the earlier successful experiments. When Denis and Emerez were called in, they "absolutely refused to make the tryal," considering the patient beyond any such assistance. Under pressure and with as great reluctance as that of the attendant doctors, they finally agreed to transfuse a small quantity of calves' blood, which they did in the morning. To the surprise of everyone, the patient rallied, his pulse became stronger, convulsions ceased, and the patient recovered consciousness, spoke, and then slept quietly. The next day, however, occurred a sinking spell. Transfusion was again tried, but without success. The patient died that night. Autopsy showed, as all the physicians had suspected, that the intestines were gangrened, and recovery would have been impossible. It seems doubtful that the transfusion contributed to the death. Inevitably, however, as Oldenburg wrote to Boyle,[40] the Baron's death caused transfusion to be "hugely disputed abroad pro and con."

It was natural that the English, who had been the pioneers in blood transfusion, as in so many other aspects of the New Science, should have resented the fact that the French outdistanced them in transfusing human beings. "The *Journal des scavans* glorieth," wrote Oldenburg resentfully in a lengthy discussion of the French experiments, "that the French have advanced the Invention so far as to trie it upon Men, before any English did it, and that with good success." The English, Oldenburg insisted, had not been backward for want of skill or

---

[40] The case was discussed by Oldenburg in a letter to Boyle, October 8, 1667 (*Works*, V, 369): "The experiment of transfusion was tried at Paris upon a baron of Sweden; but he dying his intestines were found all gangrened, so that it was not possible to recover him by any known natural means. This invention is hugely disputed abroad pro and con; of which I also received lately some specimena, both French and Latin."

lack of plan for human transfusion. Indeed, they had been ready to perform the operation some time earlier than did the French. In proof of this he quoted a letter from Dr. Edmond King, saying that he had been ready for human transfusion six months earlier and had waited for "nothing but good opportunities, and the removal of some considerations of a Moral Nature." Oldenburg enlarged upon these "considerations":

We readily grant, They [the French] were the first, we know off, that actually thus improved the Experiment; but then they must give us leave to inform them of this Truth, that the Philosophers in England had practised it long agoe upon Man, if they had not been so tender in hazarding the Life of Man (which they take so much pains for to preserve and relieve) nor so scrupulous to incurre the Penalties of the Law, which in England is more strict and nice in cases of this concernment, than those of other Nations are.[41]

Oldenburg's conclusion to the whole section is suggestive of the insistence of the Royal Society that the advancement of science was more important than the priority of any individual or nation: "But whoever this Parent may be, that is not so material, as that all that claim to this Child, should joyn together their endeavors to bring it up for the service and relief of humane life, if it be capable of it."

His passing reference to possible "Penalties of the Law" should hold our attention for a moment, since this is the first indication we have seen that experiments at Gresham College or elsewhere might prove to have legal consequences. Certainly at this time English law was paying no attention to experiments performed at the Royal Society, nor was it eight months later, when Denis in France was facing possible legal action as the result of transfusion. Apparently he had written Oldenburg to learn whether English magistrates had taken any action because of the human transfusion which had been performed. Olden-

[41] *Phil. Trans.*, II (October 21, 1667), 517–25. In Chapter IV of *De Corde*, referred to above, Lower protested vehemently and at some length about the French attempt to take away from him the credit of having performed the first transfusions.

burg wrote a prefatory note to a long letter by Denis, which will be considered presently: "After that the Author of these Tracts, having been desired from his Parisian correspondents, to inform them, Whether the Magistrate of London had forbid the use of the Transfusion of Bloud (as it was there noised) had assured them, That he never heard, that any Magistrate in England had so much as concerned themselves in this matter. . . ." We must return to France to discover the reason that, after its successful inauguration, therapy by blood transfusion was to remain dormant for well over a century.

V

Denis's letter to the Royal Society, dated January 6, 1668, described the case of "a Madman, that hath been lately cured, and restored to his wits by the means of the Transfusion." [42] This was a man about thirty-four years of age, Anthony du Mauroy Saint Amant, a *valet de chambre* when he was employed, whose "Phrensy began first of all to appear 7 or 8 years agoe," presumably as the result of an unhappy love affair. Having apparently recovered his mental health, he later married "a young Gentlewoman," but it soon became clear that his insanity was periodic, some attacks lasting as long as ten months. When Denis's attention was called to him, he had, as several times before, run away: "He got loose stark naked, and ran away streight for Paris." During lucid intervals he found occasional employment, but "his greatest divertisement during that time was, to tear the Cloaths, that were given him, to run naked abroad, and to burn in the houses where he was, whatever he could meet with." Apparently consulted by a "person of quality" who had taken an interest in the case, Denis found that the patron had "caused him to be bled in his feet, armes, and head, even 18 times, and made him bath himself 40 times." Believing that the "Bloud of a Calf by its mildness and freshness might possibly allay the heat and ebullition of his

[42] *Phil. Trans.*, II (January 6, 1668), 617–24.

Bloud," Denis took the patient to a private house and appointed as his temporary guardian that good-natured "Porter, on whom we had already practised the Transfusion 8 months agoe." The first transfusion was given at six o'clock in the evening of December 19 in the presence of "many persons of quality." [43]

Emerez opened the crural artery of the calf, took ten ounces of blood from the patient, and gave him only five or six ounces of fresh blood, in part because, as at Coga's second operation, the spectators crowded too close and interrupted the operators. Saint Amant showed sufficient improvement that Denis decided to transfuse again the following week, at which time, he reported, the patient must have received more than one whole pound of calves' blood. He fell asleep at ten that night, slept until eight in the morning, and woke apparently entirely calm and rational. When told that it was the Christmas season, he asked for his confessor, who found him sane and lucid enough that a short time afterwards—the exact time is not clear—he considered him in proper mental condition to receive communion. This letter ends happily: on January 6, 1668, the patient was sane and physically better than for a long time.

But this state of things was not to last. Toward the end of January, the periodical insanity began to reappear, thanks in part, Denis suspected, to the debauched life Saint Amant had been leading. His wife, who had returned to him, had apparently stood all she intended to stand. During his lucid period he had become suspicious of her, apparently suspected her fidelity, and accused her of trying to poison his food. The wife called in Denis and demanded another transfusion, which at first he refused to perform in spite of the fact that she had gone so far as to make all preparations, providing a calf and having

[43] Professor Harcourt Brown, in the article referred to above, gives letters from Antoine de la Poterie to Samuel de Sorbière which add numerous details. The operation was performed "en presence de M. le Comte de Frontenac, l'Abbé Bourdelot, de plusieurs scavans medicins de cette ville et dautres personnes considerables." Professor Brown identifies the physicians. Frontenac was the famous governor of Canada.

Emerez present. "His wife urged us beyond measure, to try the third Transfusion upon him, insomuch that she threatned, she would present a Petition to the Solicitor General to enjoyn us to do, which we did absolutely refuse." Threats were one thing, a woman's tears another. When Mme. Saint Amant "fell down with tears in her eyes, and by unwearied clamour engaged us not to go away," Denis reluctantly capitulated. According to his statement in this letter, all that Emerez did was to put a "pipe into the vein of the patient's arm to draw away some old bloud," but a fit of trembling seized the man and no blood came, so Emerez took the pipe out.[44] No transfusion was performed. The next evening the patient died suddenly. Denis and Emerez wished to make an immediate autopsy:

Remembering the Complaints the dead man had often made of his wives attempts to poyson him, we would gladly have open'd his Body in the presence of 7. or 8. witnesses. But she did so violently oppose it, that it was not possible for us to execute our design. We were no sooner gone, but she bestirr'd herself exceedingly, as were inform'd, to bury her husband with all speed. But being in an indigent condition, she could not compass it that day.

A distinguished—though unnamed—Paris physician, who happened to be calling at the home of a lady whom Mme. Saint Amant approached for financial aid, insisted that an autopsy be performed, as did Denis and Emerez, threatening her "that we would return next morning, and do the thing by force." Apparently she sought a carpenter to make a rude casket and to assist with burial, since when Denis and Emerez arrived, they found that she had "caused her husband to be buried an hour before day, to prevent our opening of him."

[44] *Phil. Trans.*, III (June 15, 1668), 710–13. Evidence at the preliminary hearing on April 17, 1668, is somewhat more specific: "but that almost no blood issuing neither out of the foot nor of the arme of the Patient, a Pipe was inserted, which made him cry out, though it appeared not that any blood of the Calfe had pass'd into his veins; that the Operation was given over, and that the Patient died the next night."

Inevitably rumor spread through Paris that Saint Amant had died as the result of blood transfusion.[45] Three physicians, of a more conventional school than Denis, visited Mme. Saint Amant, "importuning her by promises of a great recompence, onely to let them use her name to accuse us before a Court of Justice for having contributed to the death of her husband by the Transfusion." Since they evidently did not make it sufficiently worth her while—or because she was playing both ends to the middle—she went to Denis, offering to hold her tongue if he paid her well enough. Quite properly Denis reported the whole affair to the Lieutenant des Causes Criminelles, "who presently allowed me to inform against the Widow, and those that sollicited her."

At the preliminary hearing witnesses were brought to show that she had given her husband food not prescribed by the physicians, "strong Water to drink" and broth into which, it was said, she had introduced powders. On one occasion when her husband had accused her of attempts to poison him, she "made a shew of tasting it herself, cast it upon the ground, what she had in the spoon." On another occasion he himself gave the broth to the cat, which died not long afterward. It was proved that the powder had been arsenic. The widow was taken into custody and held for trial, which did not actually take place until late November, 1669. Because the archives of the Chatêlet were destroyed by fire in 1817, all record of the case seems to be lost with the single exception of a brief letter contributed to the *Philosophical Transactions* by "an Intelligent and Worthy English Man from Paris, to a considerable Member of the R. Society in London." [46]

The case ended in exoneration for Denis and a blaze of glory for his advocate, who, we are told, "was the Son of Mon-

[45] In his earlier letter after the successful operations (*Phil. Trans.*, II, 618), Denis said that such rumors began after the first transfusion: "Some spred a rumor that he died soon after the operation; others bore the people in hand, that he was relapsed into a greater madness, than that before."
[46] IV (December 13, 1669), 1075–77. The letter may have been to Oldenburg, as a majority of the communications from the continent usually were.

sieur de Premier President de la Moignon," a neophyte, evidently pleading his first important case. The "grand Chambre of the Parlement" was crowded by "a World of Great Persons, Men and Women." Brief though the lone extant record is, it enables us to visualize a courtroom climax to "The Strange Case of the Missing Transfusion," worthy of Perry Mason at his best. Denis, the physician, had coached his lay advocate well.[47] The defense of Jean Denis proved to be basically a defense of science, couched in language reminiscent of the Baconian program of the Royal Society, "for the advancement of learning and the relief of man's estate." The "Intelligent and Worthy English Man," then resident in Paris, had been proud to hear among the opening remarks a high tribute to the Royal Society of London, as "the Source of Noble Experiments," the leader of the world in the advancement of science. Presumably the advocate pointed out that the "invention" of transfusion had had its origin in that august body, urging his auditors not to oppose a new method because it was new. Then he made a telling point: "In Justifying the Introduction and Use of New Experiments he said, That the most precious Life to this State (viz., that of his Most Christian Majesty) had been saved by the Administration of a lately invented Emetique." It would seem that his star witnesses proved two persons he "very much gloried in . . . a Man and a Woman, present in the Audience, that received a benefit to Admiration from

[47] I have a strong suspicion that Dr. Jean Denis practically wrote his attorney's defense, since the points made in this brief account are all developed in Denis's letter to the Royal Society, particularly the one about the emetic (*Phil. Trans.*, II [December 9, 1667], 560): "There is nothing, but Experience, that is able to give the Verdict and the last Decision, especially in matters of Natural Philosophy and Physick: That a hundred years agoe, there were no Arguments wanting to prove, that Antimony or the *Vinum Emeticum* was poyson; the use of it being then forbidden by a Decree of the Faculty of Physitians; and that at this day there are no arguments wanting, to prove the contrary, and to assert; That it is a Purgative of great importance, follow'd with wonderful effects; the same Faculty having Publisht a Decree the last year, by which it permits, and even ordains the use thereof . . . and that the Recovery of many persons, and amongst them, of the Most Christian King himself, hath more conduced to convince Men of its usefulness, than all the bare Ratiocinations, that could be employ'd to defend it."

the Experiment, after they had been abandon'd by all Physitians and other helps." [48]

Although a time was set for the pleading of the "Widow Plaintiff," the general impression was that this would be a pale proceeding in comparison with the spectacular and persuasive defense. Denis was exonerated and his action justified. However, the court brought in a proviso "that for the future no Transfusion should be made upon any Human Body but by the Approbation of the Physitians of the Parisian Faculty"— a proviso which, in effect, put an end to blood transfusion, since this group was notably conservative, and many of them had opposed the experiments that had already been made.[49] La Veuve Saint Amant fades back into the shadow from which she emerged to play her part in the melodrama. As Professor Harcourt Brown said: "She drops into obscurity, while Denis, with singularly appropriate logic, takes up the development of a styptic fluid." So ends *l'affaire Saint Amant*.

VI

From melodrama I turn in conclusion to the problem which I am sure has been concerning a number of my listeners who know a great deal more than I about blood transfusion and who may well have been raising eyebrows or shaking heads at these accounts of our ancestors who performed these operations so casually. Many of their supposed successes seem, indeed, to violate all that medical science has learned of the dangers inherent in such processes. Professor Brown, in his article on Denis's transfusions in Paris, makes the point in an interesting manner:

[48] The presence of this unidentified woman bears out my belief that Denis had performed at least one more successful transfusion than the record shows. The masculine witness at the trial might have been Denis's first transfusion patient, who would now have been about eighteen. It would have been difficult for the attorney or Denis to make much of a case for the jolly porter.

[49] Keynes, p. 43, says: "As the Faculty was bitterly opposed to the whole idea, this permission was never given, and in 1670 an Arrêt forbade the practice of transfusion altogether."

The scepticism of scientists with whom I have discussed some aspects of the experiments described by Denis remains a troublesome factor in presenting this episode. The modern reader would welcome a critical discussion of the various procedures involved, their difficulties and their probable success in the light of the descriptions offered by Denis, Lower and others. If the truth of these accounts can be successfully challenged by the modern physiologist or surgeon, then the historian of science faces a serious and novel problem, the evaluation of the extent and the reason for the error or deception involved, in the light of all the circumstances under which the work was said to have been done.[50]

It seems extraordinary, indeed, that the experiments should have been as successful as the early scientists report. On the other hand, the cloud of witnesses attesting them is imposing, numbering, in England at least, some of the most distinguished names in the history of science. We have no firsthand account of the experiments Lower had made at Oxford before the one in the presence of Wallis and others, but it would be incredible that such men as Lower and Boyle had entered into a pact to lie for each other. The experiments reported by the Royal Society were performed publicly, as were most of those of Denis in France, in the presence of many witnesses, among whom were physicians and surgeons who would certainly have detected any attempt at prestidigitation or fakery. So far as animal experimentation was concerned, it is true that records report chiefly successes rather than failures, though the death of a lamb, a calf, or a dog is occasionally mentioned, even when it was not intentional, as in Lower's first operation. The foreign experiments reported to the Royal Society were, quite naturally, usually of successes. For the most part the accounts of animal transfusion that have come down to us mention the condition of the animal only immediately after the transfusion, though on a few occasions the physician or surgeon followed up the case for a time.[51] For the most part we have

[50] Page 17 n. 4.
[51] For example, an experiment was reported to the Royal Society from the Italian *Giornale de litterati*. This is given in translation in *Phil. Trans.*, III

no evidence about possible delayed reactions as a result of cross transfusion between different species of animals.

In the few cases of human transfusion the accounts are much more complete, since the physicians were naturally greatly interested in these patients. With the assistance of Dr. Geoffrey Keynes and Dr. Louis K. Diamond,[52] I shall go back over these cases and try to suggest what the effects of the transfusions actually were.

Denis's first transfusion (reportedly of eight ounces of arterial blood of a lamb), upon the young man who had had a prolonged illness, seems to have had a benign effect, although the patient mentioned "a very great heat along his arm," which might well have been a symptom of incompatible blood. However, both he and the Swedish Baron Bond, upon whom Denis's first transfusion seemed to have a benign effect, had been very ill and may illustrate the point made by Dr. Dia-

---

(May 8, 1667), 440–41. This was a transfusion from one lamb to another. The recipient, a very young animal, jumped down from the table with no sign of feebleness. Turned out to grass, it continued to grow in apparent health for seven months. When it died suddenly on January 5, 1668, autopsy showed that the stomach was "full of corrupt food. Its neck being dissected, to see what had happen'd to the vein cut through, it was found, that it had joyn'd it self to the next Muscle by some fibres, and that the upper part of that vein had a communication with the lower, by the means of a little branch, which might in some manner supply the defect of the whole trunck."

[52] I have already referred to Keynes' article on blood transfusion which I have used frequently throughout this section. Dr. Diamond is Professor of Pediatrics at the Harvard Medical School, Hematologist and Director of the Hematology Research Laboratory, Children's Hospital, and of the Blood Grouping Laboratory, Boston. I have his permission to quote a paragraph from a letter he was good enough to write me in answer to my inquiries:

"In transfusions between humans the statistical occurrence of the various blood group factors and their natural antibodies is such that the use of random donors into any recipient leads to a ⅔ chance of compatibility and only ⅓ chance of incompatibility. This would make the chance of a serious reaction in man to man transfusions not too great. Also mild hemolytic reactions would probably be ignored since only the severe and usually fatal variety might cause comment. However, transfusion of animal blood into humans would always be dangerous and could cause hemoglobinura, jaundice, anemia, and severe subjective as well as objective symptoms. The description of the more careful observers of such incidents bears this out. The only cases in which the human might receive an animal blood transfusion without immediate serious reaction would occur when the individual was quite sick and had a low antibody titre because of ill health. I'm inclined to believe that in most cases where such heterospecific transfusions were tried, the resulting signs and symptoms of incompatibility were probably ignored if not overlooked."

mond that a human being might receive animal blood without
severe reaction when he had been "quite sick and had a low
antibody titre because of ill health." It is less easy to explain
the apparent-immunity of Denis's laughing sedan-carrier, who,
Denis said (though the figure may have been exaggerated) received about twenty ounces of lamb's blood. Denis did report
that "he felt a very great heat from the Orifice of his Vein up
to his Arm-pit," but otherwise "found no indisposition in himself." He seemed in excellent health and spirits not only
immediately after the operation but some months later when
he served as warden to Saint Amant. The amount of blood received by Arthur Coga in the first operation was much less
than that administered by Denis—"eight or nine ounces by
conjecture," according to Oldenburg. He reported himself
better for the first experience and so Oldenburg reported him
to Boyle, though his account indicates at least a mild reaction
to incompatibility.[53] After the second experience Coga reported himself "very well" a week later, though he mentioned
that after the operation "he had been at first somewhat feverish." This may explain a statement made by Henry Stubbe:
"Sure I am that the Transactions report an Untruth, in saying
that Coga was ever the better for it. I am told his Arm was
strangely ill after it, and difficultly cured." [54]

In the Saint Amant case, the records, not only Denis's but
various others, are complete enough to prove that he suffered a
severe reaction. Geoffrey Keynes puts the seventeenth-century
accounts into modern language and writes:

On this occasion the blood of a calf was used, and there can be
no doubt that the patient received a considerable amount, for
he showed all the signs of receiving incompatible blood—pain
in the arm, a rapid and irregular pulse-rate, sweating, pain in
his back, vomiting, and diarrhoea; he afterwards passed urine

[53] Oldenburg reported his pulse "better" after the operation. During the night
he had two or three hours of sweating, one sign of incompatibility. He had
"three or four stools," which suggest another sign, diarrhea, though Oldenburg
says, "as he used to before."
[54] *"Plus Ultra" Reduc'd to a Non Plus,* p. 133.

that was almost black with the haemoglobin of destroyed blood-cells. In fact, he was fortunate to escape with his life.[55]

Largely as a result of the furore aroused by the Saint Amant case in Paris, and the ultraconservatism of the Faculty of Medicine, blood transfusion disappeared as a form of therapy, not to be revived until the nineteenth century.[56] In literature, as we shall see, it continued to live for some time as a subject for satire, but became so remote from human experience that we find few references to it—satiric or other—after the turn of the century. For that reason it is interesting to see that Alexander Pope coined a figure from it in the early version of his *Essay on Criticism:*

> Many are spoil'd by that pedantic throng,
> Who with great pains teach youth to reason wrong.
> Tutors, like Virtuoso's, oft inclin'd
> By strange transfusion to improve the mind,
> Draw off the sense we have, to pour in new;
> Which yet, with all their skill, they ne'er could do.[57]

Pope himself deleted the lines from his manuscript before sending it to the printer. The fact that they do not appear in any of the succeeding editions may indicate that even by 1710/1 blood transfusion had become so remote from human experience that reference to it would have lost its point.

Innocently Denis and Emerez, Lower and King, took far greater chances with the lives of their patients than they could

---

[55] Page 42.

[56] Keynes says in his article: "Nevertheless the noble experiment of transfusion in man had gained great notoriety, and some of the authors of text-books felt that they had to include an account of the procedure in their writings, though without inquiring too closely into the therapeutic results." He mentions some of these works and gives picture illustrations. Keynes also mentions the fact that in his *Zoonomia,* published in 1794, Erasmus Darwin, known to literary students through his *Botanic Garden* and other works, concerned himself with the possibilities of blood transfusion.

[57] This passage was called to my attention by George Rousseau. In the manuscript it was inserted between the present lines 25 and 26. R. M. Schmitz, (*Pope's Essay on Criticism, 1709: A Study of the Bodleian Manuscript Text,* St. Louis, 1962, p. 10) suggests a semantic reason for excision. I shall consider the matter further in a study of Pope.

possibly have realized. It may have been "beginners' luck" that no human patient died as the immediate result of blood transfusion. These men were great pioneers who took the first steps upon a long road that was to lead to one of the most important forms of medical therapy. Because of my sex, perhaps I may be forgiven for saying farewell to Lower and King, Denis and Emerez, their human patients and their animals, in familiar words of Samuel Johnson, though in much more sympathetic mood: "Sir, a woman's preaching is like a dog's walking on his hinder legs. It is not done well; but you are surprised to find it done at all."

III

"Mad Madge" and "The Wits"

ALL Pepys' references to the Royal Society and its members, both before and after his election, have indicated respect and admiration. Yet he was aware of a very different attitude among his contemporaries, which he mentioned on February 1, 1664:

> Thence to White Hall; where, in the Duke's chamber, the King came and stayed an hour or two laughing at Sir Petty, who was there about his boat; and at Gresham College in general, at which poor Petty was, I perceive, at some loss; but did argue discreetly. . . . Gresham College he mightily laughed at, for spending time only in weighing of ayre, and doing nothing else since they sat.[1]

It was, of course, inevitable that the New Science—in many ways so new and startling—should have provoked satire, suggestions of which we have already noticed in passing. Charles' courtiers were more excusable than the monarch himself for laughing at the virtuosi. Charles prided himself upon his scientific learning. He had his laboratory, which Pepys, indeed, visited later, noting on January 15, 1669: "Then down with Lord Bruncker to Sir R. Murray, into the King's little elaboratory, under his closet, a pretty place; and there saw a great many chymical glasses and things, but understood none of them. So I home and to dinner."

[1] I am purposely omitting other sentences on Sir William Petty, which I shall include later in discussing satire on his "double-bottomed boat."

Pepys has told us, as we know from other sources, that the King was interested in anatomy and that dissections of human bodies were performed for him. Although Charles II was at best a dabbler in science, he should have realized better than could Pepys the importance of "weighing the air"—those epochal experiments of Boyle and Hooke that were laying the basis for modern physics. But the King's taste led him to consort rather with "the Wits" of the day than with the scientists, and monarch, aristocrats, and gentlemen snickered behind the scenes at reports of what went on in meetings of the virtuosi. Leading members of the Society, fully aware of this, realized that the most powerful weapon turned against them was likely to be satire. Before discussing attitudes of "the Wits" toward the New Science, I shall again violate chronology in order to complete the account of Pepys and the Royal Society and begin with an occasion in 1667 which Pepys, collector of ballads as he was, was fearful might provoke broadside ballads on all sides.

I

The world of the New Science had remained as much a man's world as Bacon's travelers had found in Salomon's House in *The New Atlantis.* The first intrusion of a woman into the sacred precincts occurred in 1667 with the appearance of an extraordinary character whom Charles Lamb called "a dear favorite of mine of the last century but one, the thrice noble, chaste, and virtuous, but again somewhat fantastical and original-brained, generous Margaret of Newcastle." "Mad Madge" she was more often considered in her own day.

"The whole story of this lady is a romance, and all she do is romantick," Pepys wrote on April 11, 1667. "Her footmen in velvet coats, and herself in an antique dress, as they say." Pepys, however, freely admitted—at least to his *Diary*—that he had deliberately gone to Whitehall that day in the hope of seeing her if she made a visit to the Queen. He added: "There is

as much expectation of her coming to Court, that so people
may come to see her, as if it were the Queen of Sheba." (It is
possible that Pepys wrote "Queen of Sweden," but, with
Wheatley, I prefer the other decoding of his shorthand.) A
few days earlier Pepys, that inveterate playgoer, had gone to
see the "silly play of my Lady Newcastle's, called *The Humourous Lovers;* [2] the most silly thing that ever come upon a
stage. I was sick to see it, but yet would not but have seen it,
that I might the better understand her. . . . But she and her
Lord mightily pleased with it; and she, at the end, made her
respects to the players from her box, and did give them
thanks." [3]

No matter what Pepys' estimate of the dramatist, he spent a
good deal of time during the next few weeks trying to get a
closer view of this lioness among ladies. On April 26 he had
better luck:

Met my Lady Newcastle going with her coaches and footmen
all in velvet: herself, whom I never saw before, as I have heard
her often described, for all the town-talk is now-a-days of her
extravagancies, with her velvet-cap, her hair about her ears;
many black patches, because of pimples about her mouth;
naked-necked, without any thing about it, and a black just-au-
corps. She seemed to me a very comely woman: but I hope to
see more of her on May-day.

May Day started auspiciously for Pepys. Walking to West-
minister in the morning, he saw "many milk-maids with their
garlands upon their pails, dancing with a fiddler before
them." More exciting, he had a glimpse of Nell Gwyn, the
King's mistress, "pretty Nelly standing at her lodgings' door in
Drury-Lane in her smock sleeves and bodice," watching the
milkmaids. "She seemed a mighty pretty creature." In the afternoon he went to the theatre again, and saw another "sorry
play," at which he was pleased only by Lacy, the clown, apparently just released from prison. From the playhouse he set out

[2] Here Pepys made a mistake. This particular play was not by Margaret
Cavendish but by her husband, the Duke of Newcastle.
[3] I am conflating passages written on March 30 and on April 11.

in Sir William Penn's coach into the Park, where they encountered

a horrid dust, and number of coaches, without pleasure or order. That which we, and almost all went for, was to see my Lady Newcastle; which we could not, she being followed and crowded upon by coaches all the way she went, that nobody could come near her; only I could see she was in a large black coach, adorned with silver instead of gold, and so white curtains, and everything black and white and herself in her cap. When we had spent half an hour in the Park, we went out again weary of the dust, and despairing of seeing my Lady Newcastle, and so back the same way, thinking to have met my Lady Newcastle before she got home, but we staying by the way to drink, she got home a little before us, so we lost our labours.

On May 10, after having heard a lurid story of the killing of young Basil Fielding by his drunken brother and surveying the corpse, "a sad spectacle, and a broad wound, which makes my hand now shake to write of," Pepys "drove hard to Clerkenwell [where Newcastle House stood] thinking to have overtaken my Lady Newcastle, whom I saw before us in her coach, with 100 boys and girls running looking upon her: but I could not: and so she got home before I could come up to her." (Remember that this was almost exactly three centuries ago, you who shake your heads over decadent, modern youth, trooping after Hollywood stars or swooning over Elvis Presley and the Beatles.) But it's dogged that does it. "I will get a time to see her," said Samuel Pepys.

He did get that chance, where he had least anticipated it, at a meeting of the Royal Society at Arundel House on May 30, 1667. If Pepys could have been more regular in attending meetings—this was a busy time in his professional life—he might have saved himself his park pursuits. He found an unusually large attendance at the meeting, "in expectation of the Duchesse of Newcastle, who had desired to be invited to the Society; and was, after much debate *pro* and *con*, it seems many being against it; and we do believe the town will be full of ballads of it." Collector of ballads as he was, Pepys well knew the potency of such missiles. Ironically enough, had it not been for Pepys' graphic account of his pursuit of this elusive Pimpernel, posterity would not have paid the attention it has to the effrontery of Margaret Cavendish's inviting herself into a man's world. While Evelyn also mentioned the occasion, he did so in passing and without the detail of Pepys.[4] Evelyn's earlier comments on the Duchess, together with those of his wife, suggest that her garb and deportment impressed men more favorably than women. On April 18, 1667, Evelyn had called to pay his respects upon her arrival in London. "I was much pleasd," he commented, "with the extraordinary fancifull habit, garb, & discourse of the Dutchesse." His tone changed, however, when Mistress Evelyn accompanied him on April 27: "I went againe with my Wife to the Dutchesse of N. Castle, who received her in a kind of Transport: suitable to her extravagant humor & dresse, which was very singular." Austin Dobson, in his edition of Evelyn's *Diary*, includes a long letter from Mistress Evelyn to Dr. Bohun, from which I quote only the beginning and end:

I was surprised to find so much extravagancy and vanity in any person not confined within four walls. Her habit particular, fantastical, not unbecoming a good shape, which she may truly boast of. . . . Her discourse . . . is as airy, empty, whimsical, and rambling as her books, aiming at science, difficulties, high notions, terminating commonly in nonsense, oaths, and obscenity.[5]

Pepys' account of the Duchess' visit to the Royal Society I gave in full:

Anon comes the Duchesse with her women attending her; among others, the Ferabosco,[6] of whom so much talk is that

---

[4] Of the meeting itself Evelyn says only: "She came in greate pomp, and being received by our L. President, at the dore of our Meeting-roome, the Mace etc. carried befor him, had severall Experiments shewed before her: I . . . conducted her Grace to her Coach."

[5] London, 1908, p. 255.

[6] Wheatley says in his note that this may have been the wife or daughter either of Alfonso Ferrabosco, the younger, or of John Ferrabosco.

her lady would bid her show her face and kill the gallants. She is indeed black, and hath good black little eyes, but otherwise but a very ordinary woman I do think, but they say sings well. The Duchesse hath been a good, comely woman; but her dress is so antick, and her deportment so ordinary, that I do not like her at all, nor did I hear her say anything that was worth hearing, but that she was full of admiration, all admiration. Several fine experiments were shown her of colours, loadstones, microscopes, and of liquors: among others, of one that did, while she was there, turn a piece of roasted mutton into pure blood, which was very rare. Here was Mrs. Moore of Cambridge, whom I had not seen before, and I was glad to see her; as also a very pretty black boy that run up and down the room, somebody's child in Arundell House. After they had shown her many experiments, and she cried still she was full of admiration, she departed, being led out and in by several Lords that were there; among others Lord George Barkeley and Earl of Carlisle, and a very pretty young man, the Duke of Somerset. She gone, I by coach home.

In his use of the word "black" for the "Ferabosco" and the "pretty black boy," you must not think that Pepys was suggesting an African invasion into Arundel House; the word indicated only that they were brunette rather than blonde. In the patter of little feet, as the black boy ran from one experiment to another, I hear a sound anticipating an eighteenth-century development of "science for children," reaching a climax in one of the—to me—most obnoxious children in literature, "Tom Telescope," about whom I must talk on some other occasion. As for the Duchess, led in and out by her peers, including that "very pretty young man," how often have you and I seen her at exhibitions of various sorts, proclaiming in the idiom of the day that she is "full of admiration, all admiration."

"Several fine experiments were shown her," Pepys has said, "of colours, loadstones, microscopes, and of liquors." A pattern for exhibitions designed for distinguished visitors had been established a few years earlier when Charles II had indicated his intention of attending a meeting of the Royal Society. Lord

Brouncker, as President, had written to Christopher Wren in Oxford asking for suggestions. Wren had replied at length. One section is particularly interesting as indicating basic principles that should govern demonstrations of "science for laymen." The experiments, Wren said, "should open new light into principles of philosophy," yet they should be those "whose use and advantage is obvious, and without a lecture, and *besides may surprise with some unexpected effect* and be commendable for the ingenuity of the contrivance." [7]

A week before the visit of Margaret Cavendish on May 23 agenda of "experiments appointed for the entertainment of the Duchess of Newcastle" were drawn up and appear in the minutes:

1. Those of colours. 2. The mixing of cold liquors, which upon their infusion grow hot. 3. The swimming of bodies in the midst of water. 4. The dissolving of meat in the oil of vitriol. 5. The weighing of air in a receiver, by means of the rarefying engine. 6. The marbles exactly flattened. 7. Some magnetical experiments, and in particular that of a terrella driving away the steel-dust at its poles. 8. A good microscope.

That Boyle and Hooke had been "desired to provide and take care of" equipment and experiments was inevitable, since these were experiments earlier designed and performed by one or the other, sometimes by both. Basically the pattern established on May 23 was followed, though the unanticipated late arrival of the Duchess may have caused minor revisions.

The Royal Society rightly prided itself upon its famous instruments, which were shown to all visitors: the air pump devised by Boyle and Hooke; microscopes, probably made by Reeves, but from specifications drawn up by Hooke; and terrellas and loadstones. On the occasion of the visit Boyle demonstrated that air has weight by means of a glass receiver, weighed before and after the air had been removed. On this occasion the air proved to have weighed one ounce and seventy-one carats. There followed "the experiment of making water bubble up in the rarefying engine, by drawing out of the air; and that of making an empty bladder swell in the same engine. . . . Then the experiment of making a body swim in the middle of water." The microscopical observations were, as always, shown by Robert Hooke, probably with the assistance of some of the beautiful enlargements in the *Micrographia*, which had recently appeared. The demonstration of magnetism, a development from the work of William Gilbert,[8] was a repetition of one made by Hooke before the Society a year earlier, on April 25 and May 2, 1666:

Some experiments were made with two loadstones, one a terrella, the other of an irregular figure. Some steel-dust being scattered about them, there appeared upon the different position of the latter in respect of the former different and odd postures in the steel-dust.[9]

Boyle's final demonstration that day was of a loadstone the Society had received recently from Edward Cotton, Archdeacon of Devonshire. Its weight sixty pounds, this was then the largest loadstone in the collection. As the Duchess watched, she would have seen it producing fluctuations in the needle of the compass held about seven feet away. Perhaps as a result of the interest aroused by the Duchess' visit, Dr. Cotton sent the Society a still larger loadstone less than a month later.[10]

---

[8] The Royal Society always showed pride in the fact that the theory of magnetism was British. As John Wallis said in a later paper read before the Society: "Mr. Gilbert's Notion (of the Earths whole Body being but one Great Magnet, and lesser Magnets being so many Terrella's sympathizing with the whole) is English also. . . . And (in General) the Doctrine of Magnetism hath been more improved by our English Naturalists, than (for ought I know) by any other Nation" *(Phil. Trans.,* XXII [December 30, 1701], 1037).

[9] R. T. Gunther, *The Life and Works of Robert Hooke,* in *Early Science in Oxford* (Oxford, 1923–45), VI, 263–64. Nehemiah Grew wrote of loadstones in the earliest catalogue of the Royal Society, "Here are several, both great and small" *(Musaeum Regalis Societatis; or, A Catalogue & Description of the Natural and Artificial Rarities Belonging to the Royal Society* [London, 1681], p. 317).

[10] The loadstone shown the Duchess was described in *Phil. Trans.,* II (March 11, 1666), 423: "Of a considerable Load-stone digged out of the Ground in Devonshire. This Stone was lately sent up out of the said County, and presented

Among the various experiments performed for the noble visitor, Pepys should have been particularly interested in "those on colours," since he certainly had had the prerequisite for that course. He had been reading Boyle's *Experiments and Considerations touching Colours* only the Sunday before the Duchess' visit. That was the day he noted that, in reading the book, he laughed—a reaction that surprised us earlier, since he had elsewhere implied that he found the book difficult. Boyle would have been pleased by his laughter. He himself called this little book his "diversion," written on days "wherein having taken physic," he found himself "as unfit to speculate as unwilling to be altogether idle." He intended the book "to divert and recreate, as well as to excite." Whatever mixing of colors Boyle showed that day, he chose them well, since he had had enough experience demonstrating colors to amateurs, even to women, to make a wise choice. He said:

Ever since I did divers years ago shew some of them to a learned company of Virtuosi, so many persons of differing conditions, and even sexes, have been curious to see them, and pleased not to dislike them.[11]

Boyle undoubtedly chose from among the many possibilities experiments which stressed the element of surprise, empha-

---

to the R. Society by the Reverend Arch-deacon, Doctor Edw. Cotton, with this description, That it weighs 60 pounds; and that, though it take up no great weight, yet it moves a Needle about nine Foot distant." Boyle had either corrected Cotton's sketchy estimate or he was playing safe, since he held the compass seven rather than nine feet from the loadstone. A little less than a month after the Duchess' visit Hooke reported on June 20, 1667, that "Dr. Cotton had, according to his promise, sent to them a loadstone of about 160 pounds for a terrella, which he had chosen out of above twenty hundredweight of the same stone; and that it would move a needle at above six feet diameter," (Gunther, *Hooke*, VI, 307). It is interesting to notice that from this time on Hooke devoted a good deal of attention to trying to determine, more accurately than had been done, the force of the loadstone's attraction at varying distances.

[11] *Experiments and Considerations touching Colours*, in *Works* (London, 1744), II, 2. Professor Mintz in the article on the Duchess' visit referred to in my Preface selects from among the many spectacular transformation scenes suggested by Boyle "a sudden way of making a red colour out of two transparent liquors" by adding a drop of sulphuric acid to aniseed oil, producing a blood-red colour, and an addition of a few drops of steel solution to a glass of Rhenish wine, which turned the liquor to a beautiful green.

sized by Wren in his letter to Brouncker, sometimes associated also with an aesthetic response to the beauty of a particular color.

The demonstration of "the marbles exactly flattened" was far less spectacular than the mixing of colors. This was, however, not only one of Boyle's most famous demonstrations but was to have far-reaching consequences upon the thought of the seventeenth century. The original experiments, reported particularly in *A Defence of the Doctrine, touching the Spring and Weight of the Air*,[12] were simple enough. He had used "a pair of marbles, an inch and a half in diameter and as flat and smooth as we could get," in an attempt to explain "the cause why two very flat and smooth marbles stick so closely together that by lifting up the uppermost you may take up also the lowermost." The phenomenon had long been ascribed to one of the oldest philosophical beliefs, that "Nature abhors a vacuum," which had persisted in human thinking from Aristotle down through the ages. "I know," Boyle wrote, "the Peripateticks, and the generality of the school-philosophers, will confidently ascribe the sticking of the marbles, not to the cause we have assigned, but to nature's abhorrence and fear of a vacuum." [13] Far from acknowledging a vacuum, Boyle was attempting, as he said again and again, to "manifest, that the power or pressure of the air . . . is very great, [and] also to make some estimate (though an imperfect one) how great the power is." This very simple demonstration of the flattened marbles served not only as an important chapter in Boyle's discoveries of the nature of air, but also as one of the most powerful attacks on the Aristotelian doctrine of Nature's abhorrence of a vacuum. It brought Boyle into an arena in which he was forced to fight such "atheists" as Hobbes and such religious philosophers as Henry More.[14] Perhaps the demonstration of the flattened marbles did not interest the Duchess' ladies in

---

[12] Experiment XXXI, in *Works*, I, 110, 258.   [13] *Ibid.*, 259.
[14] Boyle mentions More's objections and quotes from his letters in *New Experiments Physico-mechanical touching the Spring of the Air*, in *Works*, III, 62, 481. Hobbes' opposition is discussed by Boyle frequently *passim*.

waiting, but she herself, I am convinced, followed it with great interest, aware as she was of many of the winds of doctrine blowing among the philosophers of the day, with whom she jealously wished to be counted.[15]

Undoubtedly the most spectacular moment of the afternoon's demonstration, which must have startled not only the ladies in waiting but the "little black boy," was that which Pepys mentioned: "one that did, while she was there, turn a piece of roasted mutton into pure blood, which was very rare." [16] Surprising though the demonstration must have seemed to the observers, it was the simplest of all the exhibitions, since the dissolution of the meat was brought about by "oil of vitriol," sulphuric acid. Supervised by those quick-change artists, Robert Hooke and Robert Boyle, the afternoon's entertainment was more than enough to justify the Duchess' reiteration that "she was full of admiration, all admiration."

After the visit to the Royal Society, Margaret Cavendish sinks into obscurity so far as Pepys is concerned, except for one passing reference. On March 18 of the following year he wrote:

Thence home, and there, in favour to my eyes, stayed at home, reading the ridiculous History of my Lord Newcastle[17] wrote by his wife, which shews her to be a mad, conceited, ridiculous woman, and he an asse to suffer her to write what she writes to him and of him.

[15] The Duchess discussed the conflicting theories of the Cartesians, the Hobbists, and the Cambridge Platonists in her *Philosophicall Letters* (London, 1664). I have mentioned some of these matters in *Conway Letters* (New Haven, 1930), pp. 177–78, 234, 237. Henry More wrote to Lady Anne Conway, a very different type of "philosophical lady," in 1664: "I am not fallen upon by one hand alone, I am spar'd by neither sexe. For I am also inform'd that the Marchionesse of Newcastle has in a large book confuted Mr Hobbs, Des Cartes, and myself" (*ibid.*, pp. 233–34).

[16] I assure the reader that there is no pun, conscious or unconscious, in the use of the word "rare." In the sense of "underdone," Pepys' generation used the word only of eggs, not of meat, and spelled it "rear."

[17] *The Life of the Thrice Noble, High, and Puissant Prince, William Cavendish, Duke of Newcastle* (London, 1667).

One could offer other interpretations of Margaret Cavendish, Duchess of Newcastle. She was an author in her own right of many other works than "silly plays." Even though some of them laughed at her, she corresponded with leading philosophers and had a sincere interest in philosophy. Before and after her visit to Gresham College she wrote on science and was sincerely interested in it. Her reading was encyclopedic. Her constant reading and writing led her physician, Sir Theodore Mayerne, to say to her husband, "If she must be a philosopher, I could wish her a Peripatetick!" [18] Exhibitionist though she was, she had real intellectual interests and abilities. What she lacked was the discipline of an education that might have brought order out of what remained chaos. As for the "ridiculous History" of her husband, while it has its faults, it has also extraordinarily vivid moments and is of real importance in the history of English literature as one of the earliest psychological biographies in our language. But for our purposes here, "mad Madge" may remain the exhibitionist Pepys thought her at a scientific meeting, after his long pursuit of her through London streets.

II

The ballads Pepys feared about the visit to the Royal Society were apparently not forthcoming, but almost from its inception members of the Society knew themselves under attack from various directions. The best proof of this is the fact that they commissioned Thomas Sprat to write their history within a half-dozen years of the existence of the organization, considering it less a history than an apologia. In a long section toward the end Sprat discussed in detail various groups that might threaten the Society because they wrongly thought that its existence threatened them. Among these were men of religion, who believed there was inevitable enmity between religion and science, and the economically-minded, who feared

[18] *Letters . . . by Margaret Newcastle* (Roxburghe Club; London, 1909), p. 4.

that the advance of the "mechanick Philosophy" would take the bread from the mouths of the laborers. Sprat rises to a climax when he comes to consider the opposition of "the Wits":

And now I hope what I have here said will prevail something with the Wits and Railleurs of this Age. . . . And indeed it has bin with respect to these terrible men, that I have made this long digression. I acknowledge that we ought to have a great dread of their power: I confess I believe that New Philosophy need not (as Caesar) fear the pale, or the melancholy, as much as the humourous, and the merry: For they perhaps, by making it ridiculous, becaus it is new, and becaus they themselves are unwilling to take pains about it, may do it more injury than all the Arguments of our severe and frowning and dogmatical Adversaries.[19]

Certainly one "ballad," written about the Royal Society in its infancy, must have had a wide circulation, since at least six manuscript copies are extant today. As three versions are available in print,[20] I shall make no effort to discuss the ballad in

[19] *The History of the Royal Society of London* (London, 1667), p. 417. In their facsimile edition of Sprat's *History* (St. Louis, 1959), pp. xxiii–xxiv, Jackson I. Cope and Harold Whitmore Jones write: "If these were the men Sprat feared, he spoke from experience. He had served for several years as chaplain to Buckingham, at whose table he had certainly met redoubtable scoffers like Samuel Butler, who seemed never to tire of goading the virtuosi." Sprat had earlier replied to Samuel Sorbière, a French virtuoso, who had visited the Royal Society in 1663, a visit reported in his *Relation d'un voyage en Angleterre* (Cologne, 1667). An English translation of the *Relation*, together with Sprat's rejoinder, may be found in *A Voyage to England, Containing Many Things Relating to the State of Learning, Religion, and Other Curiosities of That Kingdom, By Mons. Sorbière. As also Observations on the Same Voyage by Dr. Thomas Sprat, Fellow of the Royal Society, and Now Lord Bishop of Rochester* (London, 1709).

[20] The verses "In Praise of That Choice Company of Witts and Philosophers Who Meet on Wednesdays Weekly att Gresham Colledg" were first published, with copious annotations, by Dorothy Stimson, "Ballad of Gresham College," *Isis*, XVIII (1932), 103–17. I have followed her text in my quotations. Miss Stimson also included the text, with brief comment, in *Scientists and Amateurs*, pp. 56–63. F. Sherwood Taylor published another version, "An Early Satirical Poem on the Royal Society," *Notes and Records of the Royal Society of London*, V (1948), 37–46. Four copies of the ballad are in the British Museum, two among the Ashmolean MSS at Oxford. At least one copy is signed "W. G." The copy in the letterbook of Henry Power in the British Museum ascribes the poem to "Mr. Glanvill," whom C. R. Weld in his *History of the Royal Society* (London, 1848), I, 79, identifies with a "William Glanvill." Taylor suggests as

detail, but limit myself to passages that deal with members or experiments of which we have already heard. Since "each single Member hath undertooke / To shew a Trick or write a Booke," we hear in passing of John Wilkins' *Discovery of a New World in the Moon* and his book on a universal language; hear also of John Evelyn's treatise on smoke control, *Fumifugium*. We are told of the Society's interest in such practical matters as "the Philosophie of making Cloath" and in diving bells.

> For, gentlemen, 'tis no small matter,
> To make a man breath under water.

Prince Rupert's drops come in for their share of attention:

> With much adoe they shew'd the King
> To make glass Buttons turn to powder,
> If off them their tayles you doe but wring.
> How this was donne by soe small Force
> Did cost the Colledg a Month's discourse.

Some philosophers, we are told, devoted their energies to "fileing Iron into Dust" in order to test the power of their loadstones. Inevitably the satirist has his little fling at the learned man who spent his time pondering on the nature of air. Boyle's air-pump experiments at an earlier exhibition for a distinguished visitor sound very familiar to us:

> To the Danish Agent late was showne
> That where noe Ayre is, there's noe breath.

the author William Godolphin, a fellow of the Royal Society. Miss Stimson ascribes it to Joseph Glanvill, an ascription with which Jackson I. Cope agrees in *Joseph Glanvill: Anglican Apologist* (Washington University Studies; St. Louis, 1956), p. 12. I cannot share their belief. Glanvill was too eager to become a member of the Royal Society to run the risk of incurring their displeasure in any way. In addition it seems to me that Glanvill, with his feeling for sonority in style at this period, could not have written as badly and baldly, even in mock verse.

Miss Stimson dates the verses as "probably 1663," though she points out that most of the specific references are to even earlier experiments. Taylor thinks they were written between May 1 and July 24, 1661. Both datings are upon internal evidence. In some copies marginal notes or initials suggest what experimenter is referred to.

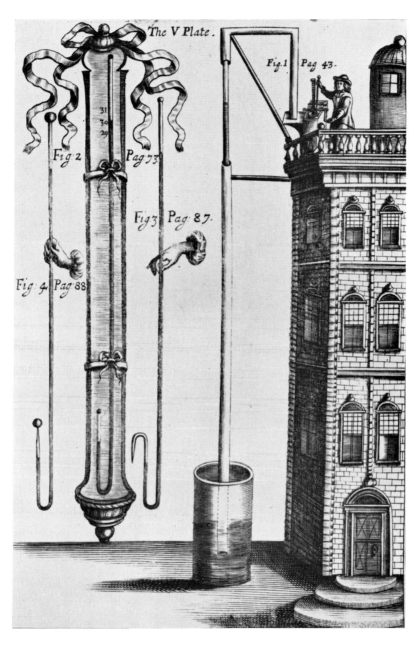

The V Plate.

Fig.1 Pag 43.

Fig 2 Pag 73.

Fig 3 Pag 87.

Fig 4 Pag 88

31
30
29

8. Weighing the air. (From Robert Boyle, *A Continuation of New Experiments Physico-Mathematical Touching the Spring and Weight of the Air,* 1669.)

9. Sir William Petty. (From a portrait attributed to Isaac Fuller; courtesy of the National Portrait Gallery, London.)

"When this portrait of Petty was bought in 1937 by the National Portrait Gallery, London, no skulls were visible. John Aubrey, the antiquary, a friend of Petty's, wrote of a portrait 'by Fuller, in his Doctor of Medicine gown, a skull in his hand,' and this led the gallery authorities to x-ray the picture. The skull was discovered under heavy overpaint, which on removal in 1958 revealed two others in the open book" (Sir Henry Hartley, ed., *The Royal Society*, p. 79; quoted by permission of the Royal Society).

> A glasse this secret did make knowne
> Wherin a Catt was put to death.
> Out of the glasse the Ayre being screwed
> Pusse dyed and ne're so much as mewed.
>
> The selfe same glasse did likewise clear
> Another secret more profound:
> That nought but Ayre unto the Eare
> Can be the medium of Sound.
> For in the glasse emptied of Ayre
> A striking watch you cannot heare.

The "Ballad of Gresham College" was written in the very early years of the Royal Society, possibly in 1661 or 1662. Another hitherto unpublished satire may have circulated in the autumn of 1663, this one dealing less with the Society as a whole than with one of its members, Sir William Petty, distinguished today in the history of economics as the father of economic statistics. The ballad is in English, though the title is Latin: "In Laudem Navis Geminae E Portu Dublinii ad Regem Carolum II^dum Missae." [21] Petty, whom we have met frequently through Pepys, was well known in his own time for many things, but most famous for the great "Down Survey" of Ireland, which he had begun in December, 1654, and finished in thirteen months—a remarkable achievement. At early meetings of the Royal Society in 1661 and 1662 Petty, as his most recent biographers write, "was the main source of information on questions of trade and industry and he was in constant request for reports on shipping, on trade, on clothing, dyeing and the history of cloth." [22] Birch's minutes indicate the great interest the Society took in Petty's invention of a new type of ship designed for easier and swifter navigation of the Irish

---

[21] Sloane MSS. 360, ff. 73–80, British Museum. I am much indebted to Professor C. A. Patrides for securing for me a photostat with great dispatch. Since the ballad is about Petty, rather than the Society in general, I treat it briefly here but discuss it in more detail in an appendix.

[22] Sir Irvine Masson and A. J. Youngson, "Sir William Petty, F.R.S.," in *The Royal Society: Its Origins and Founders*, ed. Sir Harold Hartley (London, 1960), p. 85.

Channel, a "double bottom" or "sluice bottom" boat, about which he began to communicate to the Royal Society from Ireland on November 12, 1662. Pepys' many valuable references to the vessel begin the next year, when the first ship had actually been built. On July 31, 1663, he wrote:

Mr. Grant showed me letters of Sir William Petty's, wherein he says, that his vessel which he hath built upon two keeles (a modell whereof, built for the King, he showed me) hath this month won a wager of £50 in sailing between Dublin and Holyhead with the pacquett-boat, the best ship or vessel the King hath there; and he offers to lay with any vessel in the world. It is about thirty ton in burden, and carries thirty men, with good accommodation, (as much more as any ship of her burden,) and so any vessel of this figure shall carry more men, with better accommodation by half, than any other ship. This carries also ten guns, of about five tons weight. In their coming back from Holyhead they started together, and this vessel came to Dublin by five at night, and the pacquett-boat not before eight the next morning; and when they came they did believe that this vessel had been drowned, or at least behind, not thinking she could have lived in that sea. Strange things are told of this vessel, and he concludes his letter with this position, "I only affirm that the perfection of sayling lies in my principle, finde it out who can." [23]

Pepys did not mention another prize won by the boat in this race. A committee in Ireland, appointed by the Royal Society to report on the vessel, had offered a silk flag, decorated with a gilded harp and a wreath of laurel, with the inscription, "Proemium Regalis Societatis Velociori," which was won by *Invention I,* as Petty seems to have christened the first ship.[24]

[23] In a general account of Petty, written much later on March 22, 1675, Evelyn described the earlier ship: "This vessel was flat-bottomed, of exceeding use to put into shallow ports, and ride over small depths of water. It consisted of two distinct keels cramped together with huge timbers, etc., so as that a violent stream ran between; it bare a monstrous broad sail, and he still persists that it is practicable, and of exceeding use."

[24] There is some disagreement about the names of various models. In mentioning the launching of a later ship on December 22, 1664, Pepys said, "The name, I think is Twilight, but I do not know certainly." On February 13, 1665, he mentioned the ship correctly as "Experiment," a name given it by the King at the launching. Lord Edmond Fitzmaurice, *The Life of Sir William Petty*

Novel as a "double bottom" ship was, it inevitably led to laughter as well as to great interest. If we return now to the passage from Pepys which I quoted in part at the beginning of this essay, we shall better understand sentences that deal particularly with Petty and his boat. The King, said Pepys, spent an hour or two laughing at Sir William Petty, who was there about his boat. Poor Petty was at some loss, "but did argue discreetly, and bear the unreasonable follies of the King's objections and other bystanders with great discretion; and offered to take oddes against the King's best boates; but the King would not lay, but cried him down with words only." At this point Lord Edmond Fitzmaurice in his life of his ancestor, Sir William Petty, included a sentence for which I have been able to find no authority: "He told him he would have to return to Ireland in his own ship, which he called a fantastical, bottomless, double bottomed machine." [25]

The anonymous verses are doggerel, approximately three hundred lines of iambic tetrameter couplets. They can be dated, I feel sure, as of October, 1663 because at that time Petty, hoping to win royal patronage for his invention, sent *Invention* from Dublin to Dover. The satirist assumes that the order for its departure had come from His Majesty:

> The King (God bless him) at the last,
> Did hear Fames loud, and foisting Blast,

_____

(London, 1895), p. 110, speaks of the first ship as *Experiment*. In the article by Masson and Youngson, the early model is correctly called *Invention I*.

[25] *Ibid.,* p. 112. Fitzmaurice says in his Preface that he used Wheatley's 1893 edition of Pepys' *Diary*. The sentence does not appear there nor in any other edition—Braybrooke, Wheatley, or Chandos Classics. I am indebted to George Rousseau, who checked all available nineteenth-century editions as well as more recent ones. Fitzmaurice misquotes the Wheatley edition at least twice earlier in this passage and is often inaccurate in his transcriptions, but where he found this sentence, supposedly written by Pepys, remains a mystery. I am greatly indebted to Professor William Matthews of the University of California at Los Angeles who, with Robert Latham, is editing the new edition of Pepys' *Diary*. He tells me that the sentence does not appear either in the *Diary* or in a second diary Pepys kept at the same time—a business diary called the Navy White Book. There is no entry there for February 1, 1663/4. Since Fitzmaurice had access to many Petty manuscripts, it is possible that he found the statement elsewhere and either wrongly attributed it to Pepys or, possibly, intending to paraphrase, erroneously included it between the quotation marks.

> Who being a Lover of Navigation,
> Making for Ships great preparation,
> To Ireland by post sent over,
> To have this Ship sent round to Dover.

Almost at the end of the verses the author says, "Castor, and Pollux now's on the Ocean." When the ballad was written, the ship, it would seem, had left Ireland but had not yet arrived in English waters.

Let other poets, the satirist begins his mock-heroic, spend their time in praise of ancient *Argo* or modern Armada or in arguing the "Form of which Noah his Ark made." He will celebrate a very different kind of ship,

> Whose Birth, and Breeding I'le reherse,
> To all the World in Burlesque Verse.

In the eyes of the world the creator of the "double bottom" (Petty's name is never mentioned) is acclaimed as a great inventor:

> Then first of all this famous Model,
> Sprung from a mathemattick Noddle,
> Who Honour saw, altho dimm sighted,
> And was for fair Inventions Knighted.

Like his fellows, Petty is a "projector," constantly on the watch for novel schemes and devices. His survey of Ireland completed, his restless mind sought other fields for conquest:

> Knight haveing compass'd the whole Iland
> Surveying both the Boggs and dry Land,
> Lest Witt, to work on should want Matter,
> Doth now project upon the Water. . . .
>
> In Studdy thus much Time he spent,
> Some new unheard of thing t' invent,
> A thing not found since the Creation,
> A spicke-span new Art of Navigation.

The bard now launches into an account of how the novel invention actually came into Petty's mind. Wearied by his assiduous attempts to discover a new scheme, he set out for the coffeehouse, but on his way stopped to watch some children playing with their toys. One particularly caught his attention, an ingenious lad

> Who with a Peice of Paper square,
> Form'd divers Things both neet, and rare.

From the first folds emerged "a Cock a Doodle doe," which with another fold became "a Boat with Bottom double," which the lad sent sailing in the gutter. In a trice Petty was back in his study, where he too folded paper into boats which he attempted to sail. When he "found his Paper Frigot sinking," he made a wooden model, which he sailed in a cistern. In lame mock-heroic fashion the bard describes "Knight Aeolus" puffing and huffing, raising a storm with his breath; Neptune commanding the seas

> Tho not with Trident, yet with Laddle,
> To make a rough Sea he did paddle.

Slyly the versifier suggests where credit for the great new invention actually belongs:

> Thus, first of all, as I have hinted,
> Was double Bottom Boats invented,
> And little Children did God bless 'um,
> Teach wise Philosophers their Lesson.

Since I shall later discuss the rest of the ballad, I shall at this time quote only a few lines in which Petty is specifically associated with the Royal Society:

> This Knight not full of twittle twattle
> But bookish as was Aristotle,
> To gaine the Name of a Philosopher,
> An hundred Books did turn, and toss over,
> And his cold Fingers end did blow, soe
> To become a Virtuosoe. . . .
> When he perceived his Ship would doe,
> His grand Designe he doth pursue
> By this he made another greater,

With Art, and every way compleater,
Which he presented to the Colledg
Where dwell the Witts, and Men of knowledge,
As I suppose with full Intention,
To be recorded for th' Invention.

Both these ballads are mild satire, lame and tame enough that they should not greatly have disturbed the virtuosi, no matter how widely they may have circulated. They are, however, probably suggestive of many others that have not survived or not yet been discovered.

III

The Society was to encounter more deadly missiles at the hands of the most brilliant, and most frequently quoted, satirist of the Restoration, Samuel Butler, who was as capable as Boyle and Hooke of turning meat into blood by pure vitriol. Pepys will again serve as guide, so far as *Hudibras* is concerned. On December 26, 1662, he noted: "Hither come Mr. Battersby; and we falling into a discourse of a new book of drollery in verse called Hudebras, I would needs go find it out, and met with it at the Temple, cost me 2s 6d." It would be interesting to know whether the copy Pepys purchased that day was from the advance authorized or the pirated edition. The book had been licensed on November 11, 1662, and appeared officially early in 1663. In the meantime it had been pirated. Treadway Russell Nash, in his edition of *Hudibras*, quotes an advertisement that appeared in *Mercurius Aulicus:*

There is stolen abroad a most false and imperfect copy of a poem called Hudibras, without name either of printer or bookseller, the true and perfect edition, printed by the author's original, is sold by Richard Marriott, near St. Dunstan's church, in Fleet-street, that other nameless impression is a cheat, and will but abuse the buyer, as well as the author, whose poem deserves to have fallen into better hands.[26]

[26] London, 1835, I, vii–viii. Nash spoke of this as "a ministerial newspaper, from January 1 to January 8, 1662." He was evidently using the Old Style date for 1662/3.

"False and imperfect" the pirated editon may have been, yet it had obviously made the "new book of drollery" famous before the licensed edition appeared. Pepys did not join the chorus of laughter, writing on the same day: "But when I came to read it, it is so silly an abuse of the Presbyter Knight going to the warrs, that I am ashamed of it; and by and by meeting at Mr. Townsend's at dinner, I sold it to him for 18d."

During the next month or so Pepys evidently heard the praises of *Hudibras* sung to such an extent that he decided he must have been wrong, and so on February 6, 1663, he went to a bookseller's in the Strand "and there bought Hudibras again, it being certainly some ill humour to be so against that which all the world cries up to be the example of wit; for which I am resolved once again to read him, and see whether I can find it or no." Toward the end of the year the second part of *Hudibras* was published.[27] Cannier than before, Pepys wrote on November 28, 1663:

Thence abroad to Paul's Church Yard, and there looked upon the second part of Hudibras, which I buy not, but borrow to read, to see if it be as good as the first, which the world cry so mightily up, though it hath not a good liking in me, though I had tried by twice or three times reading to bring myself to think it witty.

Even though he was no more persuaded than he had been of the merits of the book, he was enough of a collector of the "latest" that on December 10, 1663, he brought "Hudibras, both parts . . . though I cannot, I confess, see enough where the wit lies."

During the period of the *Diary* at least—it ended before the third part of *Hudibras* was published in 1678—Pepys seems never to have joined in the acclaim, though he mentioned the work on several occasions, particularly in connection with a long discourse of Sir William Petty (January 27, 1664), "one

[27] Certainly this time Pepys must have purchased the pirated edition, since the original had been licensed to Butler only on November 5, 1663, and was not published until early in 1664. The copy of *Hudibras* in the Pepys' Library is the edition of 1689, not one of these early ones.

of the most rational men that ever I heard speak with a tongue," in which he reported Petty as "saying, that in all his life these three books were the most esteemed and generally cried up for within the world—'Religio Medici,' 'Osborne's Advice to a Son,' and 'Hudibras' "—a strange juxtaposition, it seems today, of three radically dissimilar books even under as broad a term as was "wit" in the seventeenth century.[28] On another occasion (October 11, 1665) Pepys found distasteful the fact that "Mr. Seymour, one of the Commissioners for Prizes," who was at the moment "mighty high," should, in serious discourse with Brouncker and other dignitaries, "quote Hudibras, as being the book I doubt not he hath read most." But Pepys collected celebrities as assiduously as he collected books, pictures, and ballads, and the last mention of *Hudibras* in the *Diary* shows that, no matter what he thought of the poem, he recognized the fact that its author had become famous. On July 19, 1668, he took particular pains about dinner, to which he had invited a group of what might be called, in our modern jargon, "practicing artists":

Up and down in the house spent the morning getting things ready against noon, when came Mr. Cooper, Hales, Harris, Mr. Butler, that wrote Hudibras, and Mr. Cooper's cousin Jacke; and by and by comes Mr. Reeves and his wife, whom I never saw before: and there we dined: a good dinner, and company that please me mightily, being all eminent men in their way. Spent the afternoon in talk and mirth.

Samuel Cooper was the miniature painter who had painted a portrait of Mistress Pepys; John Hales (Hayls), another dis-

---

[28] Petty was apparently talking about wit as paradox: "generally like paradoxes, by some argument smartly or pleasantly urged, which takes with people that do not trouble themselves to examine the force of any argument, which pleases them in the delivery, upon a subject which they like." Unfortunately for my purposes, most of the conversation reported by Pepys had to do with Osborne, not with *Hudibras*. Francis Osborne's *Advice to a Son* was published at Oxford in two parts, the first in 1656, the second in 1658, at first anonymously. It went through five editions in two years, acclaimed by some, severely criticized and satirized by others. The advice, on Studies, Love and Marriage, Travel, Government and Religion, was sometimes aphoristic, frequently banal.

tinguished painter, had done portraits of both Pepys and his wife, as well as of Henry Harris, a well-known actor of whom we shall hear again. Mr. Reeves we have already met on various occasions. His inclusion would seem to indicate that Pepys did not consider his artists, actors, and poet as aristocrats. But let us turn from Pepys to *Hudibras*.

In Part I of *Hudibras* there was nothing to disturb the virtuosi. Attention was concentrated, as Pepys said, upon "the Presbyter Knight going to the warrs." The Society would not have taken personally a passing reference to Digby's "powder of sympathy,"

> by his side a pouch he wore,
> Replete with strange hermetic powder,
> That wounds nine miles point-blank would solder.[29]

Sir Kenelm Digby (sometime professor at Gresham College) was indeed a founding father of the Royal Society, but his fame and that of his powder had long preceded his association with the new organization. The only other allusion in Part I that could possibly be considered "scientific" is one of many comparisons made of Hudibras:

> In Mathematics he was greater
> Than Tycho Brahe, or Erra Pater:
> For he, by geometric scale,
> Could take the size of pots of ale;
> Resolve, by sines and tangents straight,
> If bread or butter wanted weight;
> And wisely tell what hour o' th' day
> The clock does strike, by Algebra.[30]

By the year 1663 even the passing trigometric reference was familiar enough and indicates no interest on Butler's part in keeping allusions to mathematics or science up to date.

The second part of *Hudibras*, appearing in 1664, is at first as

[29] I. ii. 224–26. Sir Kenelm Digby's *A Late Discourse touching the Cure of Wounds by the Powder of Sympathy* had been translated from the French by R. White, Gent., as recently as 1658, but the powder itself had been well known in the time of James I.

[30] I.i.119–26.

devoid of scientific references as was Part I. In Canto ii occurs
the brief comment on Prince Rupert's drops which I quoted
in the first essay, but the drops were so widely known by laymen that readers would not have associated the reference specifically with the Royal Society. Butler began to discover science as a satiric weapon in Canto iii, when he introduced into
his tale the character of "Sidrophel," designed, perhaps, as an
attack on William Lilly, on another astrologer, or on a com-
posite of prognosticators and almanac makers of the day. Certainly Pepys should have read this section with interest, since
he was well acquainted with some of the characters to whom
Butler specifically referred, particularly in the scene of rum-
maging Sidrophel's pockets, in which, among many odds and
ends, were found

> A copper-plate, with almanacks
> Engrav'd upon't, with other knacks,
> Of Booker's, Lilly's, Sarah Jimmers,
> And blank-schemes to discover nimmers [pickpockets].

On October 24, 1660, Pepys and Spong had paid a visit to
William Lilly "there being a club to-night among his friends."
After an evening of music in Lilly's study Pepys had gone
home with John Booker, also an astrologer, who had taken
exception to what seemed to him insincere and dishonest prognostications on Lilly's part.[31]

Sidrophel, "the Conj'rer," "this profound gymnosophist,"
proves to be, like many another of his time, one who knows a
little of everything,

> had been long t'wards mathematics,
> Optics, philosophy, and statics,

[31] The passage on Sidrophel's pockets is II.iii.1091–94. Pepys' very different
attitude toward Lilly in his only reference (June 14, 1667) may reflect merely
the general temper of the times, rather than a change in his own attitude
toward astrology after he became more familiar with science: "Thence we read
and laughed at Lilly's prophecies this month, in his Almanack this year." In
his edition of *Hudibras* (London, 1772), II, 97, Zachary Grey says that Sarah
Jimmers was an acquaintance of Lilly's, whom he called Sarah Shelborn, "a
great Speculatrix. He owns he was very familiar with her."

Magic, horoscopy, astrology,
And was old dog at physiology.

Among the many inventions and discoveries of this Universal
Philosopher, astronomy, as well as astrology, played a part:

He made an instrument to know
If the moon shine at full or no;
That would, as soon as e'er she shone, straight
Whether 'twere day or night demonstrate;
Tell what her d'ameter to an inch is,
And prove that she's not made of green cheese.
It wou'd demonstrate, that the man in
The moon's a sea mediterranean.

This is antiquated for 1663. Such satire could have been—and
often was—written nearly a half-century earlier.[32] Indeed, all
the astronomical satire in the Sidrophel passages is as passé as
was Sidrophel's telescope in the "ancient obelisk,"

From top of this there hung a rope,
To which he fasten'd telescope.

The virtuosi of the Royal Society would not have deigned to
observe through such an instrument, of a sort known to their
fathers.[33]

The antiquated equipment and astronomical opinions of
Sidrophel were undoubtedly deliberate on Butler's part. Whoever his particular victim was, he was satirizing astrologers and
consciously making his butt both ignorant and out of date so
far as astronomy was concerned. The passage on microscopy,
on the other hand, is in some ways remarkably up to date for
the year 1663, when it was written. Sidrophel's telescope may
have been passé, but his microscope seems to have been a

[32] The passages will be found II.iii.205–8; II.iii.261–68. I have discussed some
of the astronomical passages in my *Voyages to the Moon* (New York, 1948), *passim*. Any of these could have been written in the 1620's and 1630's, as parallels
were.

[33] II.iii.409–10. I shall not stop over Sidrophel's discovery of what he believed a new planet or a comet, which proved to be only a kite, flown by a
boy at night. I have quoted the passage in *Voyages to the Moon*, pp. 153–54,
and discussed it in connection with the introduction of kites into England in
the earlier seventeenth century.

modern one. Old and new are ingeniously combined in lines
297–322:

> He knew whats 'ever's to be known,
> But much more than he knew would own.
> What med'cine 'twas that Paracelsus
> Could make a man with as he tells us . . .
> Whether a pulse beat in the black
> List of a dappled louse's back;
> If systole or diastole move
> Quickest when he's in wrath, or love . . .
> How many scores a flea will jump,
> Of his own length, from head to rump.
> Which Socrates and Chaerephon
> In vain essay'd so long agone;
> Whether his snout a perfect nose is,
> And not an elephant's proboscis;
> How many diff'rent specieses
> Of maggots breed in rotten cheeses;
> And which are next of kin to those
> Engender'd in a chandler's nose;
> Or those not seen, but understood,
> That live in vinegar and wood.

Paracelsus' Homunculus, created in an alembic, had been familiar for over a century, and Socrates' flea a subject for satire
in *The Clouds* of Aristophanes. But the flea seen through magnifying glasses, as we have seen, was peculiarly a Restoration
enthusiasm.

Butler had been writing Part II during the year 1663, before the appearance of the first English book on the microscope, Henry Power's *Experimental Philosophy*, which Pepys
had read to assist him after he bought his microscope, and over
a year before the publication of Hooke's *Micrographia*. During
that year, however—and earlier—Hooke had been reporting to
the Royal Society some of his acute observations of magnified
familiar objects, which had interested all the members so
much that "Mr. Hooke was ordered to bring in one microscopical observation at each meeting." Members of the Society, and
many outsiders, were familiar with "specieses" of maggots in

cheese, since Hooke had paid particular attention to various sorts of mites, including cheese mites. They knew, too,

> Of those not seen but understood.
> That live in vinegar and wood.

Hooke had shown them denizens of many familiar objects, as well as mites in liquids, including vinegar. Many members of the general public had already learned a great deal about the magnified flea and louse. Butler was writing a little earlier than the occasion of Pepys' visit to Reeves, who showed him microscopes "which did discover a louse or a mite or sand most perfectly and largely," but he too must have had opportunity to look through microscopes at such vermin if he wished. Nevertheless, Butler's vocabulary, so far as the flea and louse are concerned, is unusually accurate for 1663:

> Whether a pulse beat in the black
> List of a dappled louse's back;
> If systole or diastole move
> Quickest when he's in wrath or love . . .
> Whether his snout a perfect nose is,
> And not an elephant's proboscis.

If Butler could have read either Power or Hooke, he would have found the "snout" of flea or louse described as a "proboscis," although both writers used the word as meaning a "probe," and not as the elephant's proboscis. He would also have found close description of the pulse beating in the back of louse or flea and of the motion of systole or diastole, which both Power and Hooke considered in detail. Neither of them raised the question

> If systole or diastole move
> Quickest when he's in wrath or love,

although both of them would have been aware of the origin of this idea, which went back to William Harvey's observations on which he based his theory of the circulation of the blood. In *De Motu Cordis* in 1628, Harvey showed that insects have hearts which reflect the circulation of the blood, as can be proved by observing their tails. In a later work, about 1649, Harvey suggested the relationship between circulation and the emotions, particularly wrath and love: "In anger the eyes are fiery and the pupils contracted; in modesty the cheeks are suffused with blushes; . . . in lust how quickly is the member distended with blood and erected." [34] It is, of course, possible that Butler had read Harvey, and very probable that Harvey's references to the effect of lust even upon animalcules had descended in risqué oral tradition, particularly among "Wits." [35]

Apparently realizing that he had struck a rich lode in satirizing at least one aspect of "science," Butler continued to develop the theme in a number of minor works, presumably

---

[34] I am greatly indebted to Professor Bernard Fleischmann of the University of Wisconsin, Milwaukee, who called my attention to these passages and sent me a copy of the Willis translation. I had written him about a passage in Meric Casaubon's "Letter to Peter du Moulin," paraphrased by R. F. Jones in *Ancients and Moderns* (rev. ed., St. Louis, 1961, p. 242), in which Casaubon mentioned "Gassendi's claim that he learned to control his passions by observing how all the blood of a louse when angered ran into its tail." For background of that tale, he sent me the Harvey passages, from Robert Willis' edition of Harvey's *Works* (London, 1847, repr. Ann Arbor, Mich., 1943). On p. 29 Willis translates *De Motu Cordis*:

"I have also observed that almost all animals have truly a heart, not the larger creatures only . . . nay, even in wasps, hornets, and flies, I have, with the aid of a magnifying glass, and at the upper part of what is called the tail, both seen the heart pulsating myself, and shown it to many others."

The passage I have quoted in the text is from Harvey's "Second Disquisition to John Riolan, Junr. in Which Many Objections to the Circulation of the Blood Are Refuted," c. 1649, in *Works*, p. 129).

[35] One recently published book would have given Butler material: Edward Topsell, *The History of Four-footed Beasts and Serpents* (London, 1658). As an appendix Topsell added a very remarkable book, *The Theater of Insects, or Lesser Living Creatures: As Bees, Flies, Caterpillars, Spiders, Worms, etc. A Most Elaborate Work. By T. Moffet, Dr. of Physick*, pp. 889–1130. The work seems to have been intended for publication shortly after the death of the author. The preface by Sir Theodore Mayerne is dated 1634. Mayerne says that Moffet made the observations with the naked eye, though it is possible, of course, that he had some form of magnifying glass. Even with that, his sight must have been remarkably acute. Both *The History of Four-footed Beasts* and *The Theater of Insects* are profusely illustrated, and both contain sections on fleas and lice, though Moffet goes into more detail than does Topsell. Many varieties of lice are described by Moffet on pp. 1093 ff. In neither of these, however, have I found the somewhat technical vocabulary used by Butler, nor any suggestion of the effect of emotions upon circulation.

written during the years that elapsed between Parts II and III of *Hudibras*.[36] Some of these, I feel sure, circulated in manuscript among the Wits at taverns and coffeehouses. The latter, which had begun to appear sporadically toward the end of the Commonwealth period in the university towns and in London, served on the one hand to further the New Science, on the other to attempt pulverization of the virtuosi by laughter. As Pepys has shown us, members of the Royal Society had their favorite taverns and coffeehouses where they gathered for conversation before or after their meetings. "Penny Universities" the coffeehouses proved to be, where intelligent men of limited means might share in intellectual conversation or upon occasion hear informal or formal lectures on subjects connected with science and philosophy. While still a student at Oxford, Christopher Wren had been one of several later members of the Royal Society who had persuaded an apothecary to sell "coffey publickly in House against all Soules Colledge" as early as 1655. At Cambridge attendance of students at coffeehouses increased to such an extent that a statute adopted on November 9, 1664, directed that "all *in statu pupillari* that shall go to coffee-houses without their tutors leave should be punished according to the statute for haunters of taverns and ale-houses." A Cambridge don, John Houghton, defending coffeehouses as places where students might learn more than in lecture halls, quoted "one who he said was a former member of the Royal Society who had ventured to compare a coffee-house

---

[36] In this section, in common with all students of Butler, I face two problems I do not pretend to have solved: the authenticity and the dating of minor works attributed to Butler. Butler's biography has never been fully established. The article in the *DNB* is often incorrect. To some extent it has been corrected by René Lamar in *Revue anglo-américaine*, I (1924), 213–27, by E. S. de Beer in *Review of English Studies*, III (1928), 159–66, and in *TLS* (1940), p. 327. But lacunae still remain. In spite of the fact that the phrase "the Genuine Remains of Mr. Samuel Butler" has been repeated since R. Thyer used it in 1759, there is still a question how genuine some of these verses are. For my purposes here, attribution to Butler is not important, since I am concerned with satires on the Royal Society rather than with Butler *per se*. The question of dating, however, is very important to me, since I am particularly interested in trying to find why Sprat made such a point of "the Wits" as early as he did.

with the University itself. 'That worthy,' said Houghton, had a
very good esteem to the Universities and they for him,' and often expressed the opinion 'that coffee-houses had improved useful knowledge, as much as they [the universities] have,' and
spake in no way of sleight to them neither." The Cambridge
fellow added: "They are both best, but I must confess, that he
who has been well-educated in the schools, is the fittest man to
make good use of coffee-houses." [37]

The term "Wits' Coffee House," used by various modern
historians, has sometimes been attributed to Pepys. He did not
actually use the phrase, although it is implicit in his account of
his first visit to a coffeehouse familiar to modern literary students. At the time Pepys mentioned it, it was "the Rose." Later, from the proprietor's name, it became "Will's." Here a
chair was reserved for Dryden throughout the year. Pepys
wrote on February 3, 1664:

In Covent Garden to-night, going to fetch home my wife, I
stopped at the great Coffee-House there, where I never was before; where Dryden the poet (I knew at Cambridge) and all
the wits of the town, and Harris the player, and Mr. Hoole of
our College [Magdalene]. And had I time then, or could at
other times, it will be good coming thither, for there, I perceive, is very witty and pleasant discourse.

I suspect that, among the Wits who passed exaggerated tales of
the Royal Society from mouth to mouth or hand to hand, was
this Henry Harris, the actor, later a fellow guest with Samuel

---

[37] The statments about coffeehouses at Cambridge are quoted from Aytoun
Ellis, *The Penny Universities* (London, 1956), pp. 28–29. He does not identify
"the former member of the Royal Society" or give the source of his quotation.
Other illustrative detail, including the quotation about the early establishment
of coffeehouses at All Souls College will be found in Chapter II, "The Royal
Society and the Scientists of England," in Lewis S. Feuer, *The Scientific Intellectual* (New York and London, 1963), particularly pp. 52–56. In the most
recent study of coffeehouses, B. Lillywhite says (*London Coffee Houses: A Reference Book of Coffee Houses of the Seventeenth, Eighteenth, and Nineteenth
Centuries* [London, 1963], p. 662) that "Wits' Coffee House" was a "fleeting
title, or description, given to, or acquired by various coffee houses at different
times. Its use seems to have been determined by the character of the frequent-
ers of a house, and by the ebb and flow of popularity." Lillywhite adds that
Pepys may have originated the term in 1663–64.

Butler at Pepys' dinner. Not long before that occasion Pepys wrote, on April 29, 1668, of a visit to the Duke of York's playhouse, where he saw Harris in *Love in a Tub.* After the play, he stopped in Harris' dressing room, which he had never visited before, "and there I observe much company come to him, and the Witts, to talk, after the play is done, and to assign meetings."

Butler suggests that laughter at Gresham College had become a staple of coffeehouses as early as 1663, when he spoke in the "Sidrophel" canto of *Hudibras* (II.iii.809–10) of

> Those wholesale Critics, that in coffee-
> Houses cry down all philosophy.

In "The Elephant in the Moon" he put a similar couplet into the mouth of one of his speakers, whom I shall later identify as Robert Boyle (ll. 205–6):

> Nor shall our ablest virtuosos
> Prove argument for coffee-houses.

Whatever the Wits thought of Boyle, he had a lower opinion of them, which he did not disguise in a lengthy "digression on wit and the wits" in *Some Considerations touching the Style of the Holy Scriptures,* published in 1661.[38] To a man of his cultivation in the humanities as well as the sciences, with Latin as familiar as English, they seemed an ignorant and irreverent crew:

They speak arrogantly and censoriously both of God and men, whilst themselves oftentimes understand no tongue but their mother's; and are strangers enough to rhetorick, nor know the difference betwixt a trope and a figure, betwixt a prosopopoeia and a metaphor, or betwixt a climax and a metonymy.

To this kind of "trillo"—as Butler characterized Boyle's style— Butler might have been replying in his couplet on *Hudibras* (I.i.81–82):

[38] In *Works,* II, 125–32.

> For Rhetoric he could not ope
> His mouth, but out there flew a trope.

Boyle could be succinct, as when he wrote of the coffeehouse Wits: "Some of these gallants by their tavern songs use the muses like anchovies, only to entice men to drink."

The speakers in "The Elephant in the Moon," as we shall see, were very conscious of the discredit suffered by the virtuosi (ll. 393–96):

> we're cry'd down,
> And made the talk of all the Town,
> That rants and swears, for all our great
> Attempts, we have done nothing yet.

The mirth of the Wits must have seemed to the virtuosi uncomfortably close to that of the King in the Duke's chamber in 1664, laughing at Gresham College for weighing the air and doing nothing else since they sat. *Hudibras,* they knew, was very familiar to His Majesty. The provenance of numerous volumes shows that the King presented copies to his favorites. Butler himself said:

> He never ate, nor drank, nor slept,
> But "Hudibras" still near him kept;
> Nor would he go to church or so,
> But "Hudibras" must with him go.[39]

If Butler's shorter verse and prose satires were circulating in manuscript in the coffeehouses, the Royal Society could be fairly certain that copies had been or promptly would be made for the King and his court.

It seems probable that the first of Butler's minor works satirizing the Royal Society was a sequel to the third canto of Part II of *Hudibras,* in which Butler continued the use of the

---

[39] Quoted in the article on Samuel Butler in the *DNB.* I have not been able to find the passage. Anthony à Wood says in *Athenae Oxonienses* (London, 1691, I, 326): "His Majestie had a respect for him, and the more, for his poem called *Hudibras;* the first part of which . . . was not only taken into his Majesties hands, and read by him with great delight, but also by all Courtiers, loyal Scholars and Gentlemen, to the great profit of the Author and Bookseller."

name "Sidrophel"—"An Heroical Epistle of Hudibras to Sidrophel." There is no evidence for the date of composition. Certainly this Sidrophel is not Lilly or any other astrologer or almanac maker, but one of the Gresham College virtuosi, an identification implied in ll. 81–88:

> Hence 'tis because ye've gain'd o' th' college
> A quarter share, at most, of knowledge,
> And brought in none, but spent repute,
> Y' assume a pow'r as absolute
> To judge, and censure, and controll,
> As if you were the sole sir Poll,
> And saucily pretend to know
> More than your dividend comes to.

In his edition of *The Genuine Remains* [40] of Butler, in which many of the minor works first appeared, R. Thyer said that he could "assure the Reader, upon the Poet's own Authority, that the character of Sidrophel was intended for a Picture of sir Paul Neal." Sir Paul Neile, the eldest son of Richard Neile, Archbishop of York, and father of William Neile, a brilliant mathematician who died young, is better known through his family than in his own right. He contributed nothing to the history of science, but he was active in the affairs of the Society and was an intelligent amateur astronomer, noted particularly for his remarkable collection of telescopes, construction of which he supervised. Evelyn mentions that on the evening of May 3, 1661, he was present when the King, with Neile, Brouncker, and others, observed Saturn through the thirty-five-foot telescope which had been a gift of

[40] *The Genuine Remains in Verse and Prose of Mr. Samuel Butler . . . , with Notes by R. Thyer, Keeper of the Public Library at Manchester* (London, 1759), p. 1. Treadway Russell Nash inclined to this identification, although he pointed out (II, 93–94) that Zachary Grey had opposed it on the basis of ll. 91–92:
> No, tho' ye've purchas'd to your name
> In history, so great a fame,
holding that "Lilly's fame abroad was indisuptable," while Neile's was not. He is correct enough in the last phrase, since Sir Paul Neile is the only founder of the Royal Society whose name does not appear in the *DNB*. Little is known of him in comparison with the others. The best account is the recent one by C. J. Ronan and Sir Harold Hartley in *The Royal Society*, ed. Hartley, 159–66.

Neile to His Majesty. He also presented to the Royal Society a thirty-five-foot instrument, through which Hooke made many observations, and in September, 1664, made them a gift of a fifty-foot instrument. Neile seems to have been *persona grata* to Charles, since Birch's *History* and other sources indicate that he frequently acted as liaison between the King and the Society. Zachary Grey, the early editor of *Hudibras,* is responsible for the story that Neile had offended Butler by asserting that he was not the author of *Hudibras.* Whether this is true or not, I think there is little doubt that "Sir Poll" was intended for Sir Paul Neile, since I find that he is the only "Sir Paul" listed in the membership of the Royal Society until 1674.[41]

Butler makes no attempt at verisimilitude, so far as "Sir Poll's" scientific accomplishments and inventions are concerned, as, indeed, he could not. He is content with attributing to him many of the enthusiasms of the Society. He had evidently been the inventor of the speaking trumpet, a companion to the otacousticon, and to what end?

> your new nick-nam'd old invention
> To try green-hastings with an engine.

It will be used—if at all—by costers selling fruit or vegetables on the street. He is presumably the father of the "whimsy'd chariots," of which we have already heard, as well as of blood transfusion, a brief passage on which I shall mention later. Like Hooke and Boyle, he had studied the nature of sound and sought it in one of the few places those experts had not investigated, rotten eggs. He seems to have been a specialist in agriculture, horticulture, and particularly dendrology:

---

[41] Thyer's references to Grey are to an edition published in 1744, which has not been available to me. I have used a later edition, *Hudibras,* ed. Zachary Grey (2 vols., London, 1772). Grey's attribution to Lilly rather than Neile will be found II, 111 n. The reference to Neile's charge against Butler is II, 105. Not only was Sir Paul Neile the only "Sir Paul," but with one exception he was the only "Paul" until Paul Rycaut was admitted in 1666. Rycaut was later knighted but not for some time.

> Your sev'ral new-found remedies,
> Of curing wounds and scabs in trees:
> Your arts of fluxing them for claps,
> And purging their infected saps;
> Recovering shankers, crystallines,
> And nodes and blotches in their reins.[42]

The Royal Society was, of course, greatly interested in anything that would improve the cultivation of trees and plants. It maintained a standing "Committee for considering of Agriculture and the History and Improvement thereof," through which numerous reports were received.[43] Of all their members, no one was more interested in such matters than John Evelyn, who on February 16, 1664, had presented to them his *Sylva; or, A Discourse on Forest Trees.* It is all very well, the satirist is saying in effect, for you to experiment with the wood of trees, but what about your own wooden head? All these scientific discoveries

> Have no effect to operate
> Upon that duller block, your pate.

In common with all his fellows at Gresham College, "Sir Poll" boasts of his accomplishments and grossly exaggerates:

> now your talent's so well-known,
> For having all belief out-grown,
> That ev'ry strange prodigious tale
> Is measur'd by your German scale,
> By which the virtuosi try
> The magnitude of ev'ry lie.
> Cast up to what it does amount,
> And place the bigg'st to your account.

Since a German mile was equivalent to approximately four English miles, one can readily tell the proportionate discount a listener should attach to the tall tales of virtuosi who

---

[42] The speaking trumpet is mentioned in ll. 21–22; the chariot in ll. 57–58; sound from rotten eggs, l. 46; "curing wounds and scabs in trees," ll. 48–52.

[43] An extensive report from this committee was published in *Phil. Trans.,* I, (July 3, 1665), 91–94.

Unriddle all that mankind knows
With solid bending of your brows.
All arts and sciences advance,
With screwing of your countenance,
And with a penetrating eye,
Into th' abstrusest learning pry.[44]

Treadway Nash was correct, I think, in feeling that this set
of verses is inferior to much of Butler's work. That criticism
Butler himself would seem to have made of another brief
poem, the "Satire upon the Royal Society," since he left it a
fragment. If Butler intended this as specific rather than general satire, his target may well have been Robert Hooke:

A learned man, whom once-a-week
A hundred virtuosos seek,
And like an oracle apply to,
T' ask questions, and admire, and lie to.

Certainly there is no one of the inventions and experiments
Butler exaggerates there that could not be attributed to that
versatile center of the Society. Hooke, as we remember, had
lectured on

Which way the dreadful comet went
In sixty-four, and what it meant

although Butler, like many of his contemporaries, was more
interested in possible influences and implications of the comet
in politics, wars, plagues, than in its periodicity. If Hooke had
not actually solved such problems as

Why currents turn in seas of ice
Some thrice a-day, and some but twice?
And why the tides at night and noon,
Court, like Caligula, the Moon?

he had his theories on tides and on various occasions showed
much interest in Arctic regions. We know that he had spent a
good deal of time on

the strange magnetic cause
The steel or loadstone's drawn or draws.

[44] The first passage will be found in ll. 93–100, the second, ll. 67–72.

I am sure that Samuel Pepys, after his many conversations with Hooke on the nature of sound, would have agreed that Hooke (if anyone) could

> in the braying of an ass
> Find out the treble and the bass;
> If mares neigh alto, and a cow
> A double diapason low.

But these and the other theories Butler casually tosses out were not peculiar to Hooke, even though he and Boyle were Universal Philosophers. The satire here is, I think, generalized rather than personal. It is upon the "learned speculations . . . and constant occupations" of the virtuosi as a group

> To measure wind, and weigh the air,
> And turn a circle to a square;
> To make a powder of the sun,
> By which all doctors should b' undone;
> To find the north-west passage out,
> Although the farthest way about.

Neither "Hudibras to Sidrophel" nor this fragment has the brilliance that sparkles through Butler's most extensive and best satire on the Royal Society, "The Elephant in the Moon," which remained unpublished until Thyer included it in *The Genuine Remains* in 1759. Butler himself seems to have liked this so well that he made two versions, one in octosyllabics, the other in long-verse. "That this was compos'd after the other," said Thyer, "is manifest from its being wrote opposite to it upon a vacant Part of the same Paper." [45] Of all Butler's shorter poems, this has been most frequently republished, quoted, and discussed. I shall therefore merely sum up the story, familiar to many readers, and devote myself chiefly to an attempt to identify the scientists who are satirized. Upon an

[45] *Genuine Remains*, I, 26. I follow the text given by R. B. Johnson, *Poetical Works of Samuel Butler* (Aldine British Poets; London, 1893), II, 123–56, because for this poem it is more practical for teacher and student than the Lamar edition. In the long-verse version Johnson gives all added passages, in inverted commas, in the text. Lamar gives some of them in end notes and some not at all.

evening when the "radiant light and influence" of the moon
were at their height, a group of virtuosi turned their telescope
to the heavens. In the moon, which they took for granted was
both inhabitable and inhabited, they found that war was going
on between two parties. Not only that. As they observed more
closely, they discovered that one group of the fighting forces
made use of elephants. In great excitement, they agreed "to
draw an exact Narrative" of what their eyes had seen:

> they all with one consent
> Agreed to draw up th' Instrument,
> And, for the general satisfaction,
> To print it in the next "Transaction."

While they sat at their tables, industriously compiling their
extraordinary report,

> The foot-boys, for diversion too,
> As having nothing else to do,
> Seeing the telescope at leisure,
> Turn'd virtuosoes for their pleasure.

Their minds unclouded by learned hypotheses, simple straight-
forward observation showed one of them promptly what had
happened:

> For he had scarce apply'd his eye
> To th' engine, but immediately
> He found a mouse was gotten in
> The hollow tube, and shut between
> The two glass windows in restraint,
> Was swell'd into an Elephant,
> And prov'd the virtuous occasion,
> Of all this learned dissertation.

Obviously the footboys had not cleaned the telescope for a
long time, since, when it was dismantled, not only was the
mouse discovered, but the lunar armies proved to be

> prodigious swarms
> Of flies and gnats, like men in arms.

And the moral of all this, Butler says, is

as a mountain heretofore
Was great with child, they say, and bore
A silly mouse; this mouse, as strange,
Brought forth a mountain in exchange.

In his edition of *The Genuine Remains* Thyer said that
"The Elephant in the Moon"

was founded upon a Fact mentioned in a Note upon Hudibras'
Heroical Epistle to Sidrophel, where the Annotator observing,
that one Sir Paul Neale, a conceited Virtuoso, and Member of
the Royal Society was probably characterized under the person
of Sidrophel, adds— "This was the Gentleman who, I am told,
made a great Discovery of an Elephant in the Moon, which
upon Examination, proved to be no other than a Mouse,
which had mistaken its Way and got into the Telescope.[46]

This story, I suspect, was either apochryphal or an exaggeration of the Wits, based upon Neile's extensive collection of
telescopes. Had Butler known it at the time, would he not
have used that as his point of departure in "Hudibras to
Sidrophel, "which might thereby have become "The Elephant
in the Moon"?

Throughout the satire Butler had in mind one of the most
widely known works of "popular science" of the seventeenth
century, written by John Wilkins, who was to be not only one
of the founders of the Royal Society but, as Warden of Wad-
ham College, 1648–1659, the center of the Philosophical
Society of Oxford, various members of which were cofounders
of the London Society. In his youth (he had been only twenty-
four when the first edition appeared) Wilkins had written a
book of which the long title tells the substance: *A Discourse
concerning a New World and Another Planet: The First
Book, The Discovery of a New World; or, A Discourse Tending to Prove, that 'Tis Probable There May Be Another Hab-*

---

[46] I, 1. Zachary Grey tells the story about Neile in his Hudibras, II, 105. So
far as I can see, no one of the speakers in "The Elephant" suggests Sir Paul
Neile, in spite of the close parallel between a passage I have quoted from
"Hudibras to Sidrophel" and 11. 453–58 in "The Elephant." Such repetition is
common in Butler and usually non-significant.

itable World in the Moone, which appeared in 1638. Within two years a second and a third edition were called for. To his own embarrassment Wilkins became one of the lions of the day. Since we have become acquainted with Margaret Cavendish, I may stop for a story about her. Meeting Wilkins, she raised a problem about his lunar astronauts: "But where, Sir, shall they be lodged, since you confess there are no inns on the way?" Wilkins suavely replied: "Surely, your Grace, who has written so many romances, will not refuse my mariners rest and refreshment in one of your many castles in the air!"

Among the theories Wilkins had propounded was the possibility of colonizing the moon, which, according to him, Kepler also anticipated. Wilkins' little book shows a great deal of influence from Kepler's *Somnium,* which, although written much earlier, had been published only four years before Wilkins'. With Kepler, Wilkins posited the possibility that the moon was divided into two zones, Privolva and Subvolva, which Butler exaggerated into the two warring armies of Privolvans and Subvolvans. Now let us go back to the opening lines of "The Elephant in the Moon":

> A learn'd Society of late,
> The glory of a foreign state,
> Agreed upon a summer's night,
> To search the Moon by her own light;
> To Take an invent'ry of all
> Her real estate and personal;
> And make an accurate survey
> Of all her lands, and how they lay,
> As true as that of Ireland, where
> The sly surveyors stole a shire:
> T' observe her country, how 'twas planted,
> With what sh' abounded most, or wanted;
> And make the proper'st observations
> For settling of new plantations,
> If the Society should incline
> T' attempt so glorious a design. . . .
> Impatient who should have the honour
> To plant an ensign first upon her.

Basically this is Wilkins, yet we catch another overtone; in lines 5–10 Butler reminds us of Sir William Petty's "Down Survey" of Ireland and suggests that he and the other surveyors "stole a shire" of that country.

The first speaking character is naturally Lord Brouncker, the perennial President of the Royal Society, "one, who for his deep belief was virtuoso then in chief." In the octosyllabic version, the second speaker would seem to have been Boyle:

> a great philosopher,
> Admir'd and famous far and near,
> As one of singular invention,
> But universal comprehension,
> Apply'd one eye, and half a nose,
> Unto the optic engine close:
> For he had lately undertook
> To prove, and publish in a book,
> That men, whose nat'ral eyes are out,
> May, by more pow'rful art, be brought
> To see with th' empty holes, as plain
> As if their eyes were in again;
> And if they chanc'd to fail of those,
> To make an optic of a nose,
> As clearly, it may, by those that wear
> But spectacles, be made appear
> By which both senses being united,
> Does render them much better sighted.

Among his many other achievements Boyle was recognized as an authority upon sight. Indeed, we remember that Pepys sought him out in order to consult him about his eyes, as did many others. Among recipes he left were numerous ones dealing with eye trouble of various sorts. While he naturally never went to such extremes as Butler suggests, he was greatly interested in accounts of individuals in whom another sense—such as smell or touch—seemed to compensate for loss of sight, as in the case of a Dutch John Vermassen who could recognize colors by touch. Retold by Boyle at length, this was picked up from him by Swift in the third book of *Gulliver's Travels*.[47]

[47] The account is referred to by Boyle more than once; the extended account

In the long-verse version, however, Butler introduced into the middle of the corresponding passage lines which at first seem to have no connection with the rest of the section. After "universal comprehension" he wrote:

> By which he had compos'd a pedler's jargon,
> For all the world to learn, and use in bargain,
> An universal canting idiom,
> To understand the swinging pendulum
> And to communicate, in all designs,
> With th' Eastern virtuoso Mandarines.

Thyer [48] associates these lines with a similar passage in the Sidrophel canto of *Hudibras* (II.iii.1021 ff.), after which Thyer had quoted a paragraph from Butler's notes, which he repeats here:

The Device of the Vibration of a Pendulum, was intended to settle a certain Measure of Ells and Yards, &c. (that should have its Foundation in Nature) all the World over: For by swinging a Weight at the End of a String and calculating (by the Motion of the Sun, or any Star) how long the Vibration would last, in Proportion to the Length of the String, and Weight of the Pendulum; they thought to reduce it back again, and from any Part of Time compute the exact Length of any String that must necessarily vibrate, into so much Space of Time: So that if a Man should ask in China for a Quarter of an Hour of Sattin or Taffata, they would know perfectly what it meant; and all Mankind learn a new Way to measure Things no more by the Yard, Foot, or Inch, but by the Hour, Quarter, or Minute.

Here Butler was satirizing a scheme for a universal language which had been proposed by John Wilkins in an early work, *Mercury: The Secret and Swift Messenger* (1641), upon a

---

is in *Works*, II, 14b. For Swift's use of the tale, see Nora M. Mohler and Marjorie Hope Nicolson, "The Scientific Background of Swift's *Voyage to Laputa*," *Science and Imagination* (Ithaca, N.Y., 1956), pp. 143–44. Boyle's interest in such matters may be seen in "Some Uncommon Observations about Vitiated Sight," appended to *A Disquisition about the Final Causes of Natural Things* (in *Works*, IV, 551–56) and in a lengthy section on the human eye in the second part of the *Christian Virtuoso* (*ibid.*, V, 696–700).

[48] *Genuine Remains*, I, 30–31.

much more elaborate version of which Wilkins continued to work, as was well known, since, far from making any secret of his design, he was constantly on the watch for assistance in various kinds of vocabulary. Wilkins' scheme had been familiar to the anonymous author of the "Ballad of Gresham College," who wrote as early as 1662–63:

> A Doctor counted very able
> Designes that all Mankynd converse shall,
> Spite o' th' confusion was att Babell,
> By Character call'd Universall.
> How long this Character will be learning,
> That truly passeth my discerning.

Perhaps because of all the mechanical contrivances Wilkins had described in another book, *Mathematical Magic, or, The Wonders That May Be Performed by Mechanicall Geometry* (1648), Butler simply mechanized the universal language, turning it into one of the many pendulums in the possession of the Royal Society, for which Boyle and Hooke were particularly responsible.[49] His technique here is somewhat similar to that of Swift, whose satire on the Royal Society often resembles Butler's, when he described the group in Laputa who never used language for communication but carried with them baskets filled with geometrical figures, which they held up to one another.

Pepys first mentioned Wilkins' scheme for a universal language when he wrote, on January 11, 1666, that Wilkins was just finishing the work. On June 4 of that year, proud to be of service, he lent Wilkins "some of my tables of naval matters, the name of rigging and the timbers about a ship, in order to Dr. Wilkins' book coming out about the Universal Language." Publication had been scheduled for the autumn of 1666. Wil-

---

[49] Butler does not seem to have had a particular pendulum in mind. One group of pendulum experiments, which must have attracted general attention, was carried on from the steeple of St. Paul's by Hooke and others in August and September, 1664. Hooke mentions among his collaborators on those occasions Sir Robert Moray and John Wilkins (Gunther, *Hooke*, VI, 189, 192, and *passim;* Boyle, *Works*, V, 306–7, 534).

kins' dedication to Lord Brouncker of the volume published in 1668, tells the story:

When this work was done in Writing, and the Impression of it well nigh finished, it hapned (among many better things) to be burned in the late dreadfull Fire; by which all that was Printed (excepting only two Copies) and a great part of the unprinted Original was destroy'd.

When the book finally appeared, Pepys bought a copy on May 15, 1668, and on various occasions noted that his wife or boy was reading it aloud to him. His last mention of the book occurs on October 18 of the same year, shortly after Wilkins had been ordained Bishop of Chester. Pepys, calling upon him, "had most excellent discourse about his Book of the Reall Character."

The next speaking character in "The Elephant in the Moon" began somewhat vaguely in Butler's mind, since in the octosyllabic version he wrote only:

> Another, of as great renown,
> And solid judgment, in the Moon,
> That understood her various soils,
> And which produced best [gennet-moyles] [50]
> And in the register of fame
> Had entered his long-living name.

In the long-verse version generalized satire becomes specific, and Butler's barb is turned upon John Evelyn:

> Another sophist, but of less renown,
> Though longer observation of the Moon,
> That understood the diff'rence of her soils,
> And which produc'd the fairest gennet-moyles,

[50] While I frequently find it difficult to determine some of Thyer's editorial policies, I think that "gennet-moyles" was not in the octosyllabic version; Thyer italicizes it, as he does all the words Butler added in long-verse, in the passage, ll. 113–18. Thyer identified this speaker with Evelyn on the basis of the "gennet-moyles" and *Pomona*. He does not interpret the passage on St. Paul's and paving London. With the exception of Lord Brouncker, Thyer does not identify any of the other speakers. I may add that much later, after Butler's death, Evelyn did publish a paper on soil, his *Terra*, based upon a paper read to the Royal Society in April, 1675.

> But for an unpaid weekly shilling's pension,
> Had fin'd for wit, and judgment, and invention.

We notice immediately that in the second version, the character has been demoted from "great renown" to "less renown," and that the suggestion that his "long-living name" has been entered in fame's register has disappeared. This is entirely legitimate, since, charter member though he was and important in the foundation of the Society, Evelyn made no pretense to authority in any branch of science. He knew a great deal about the "diff'rence of soils" on earth, if not in the moon, as he has showed in *Sylva*. More important for the interpretation of "gennet-moyles"—a variety of apple—is the fact that he particularly mentioned them in a shorter supplement to the *Sylva* called *Pomona,* to which Evelyn added an Appendix on "Cyder," a history of apples and their culture. If the lines on the "weekly shilling's pension" suggest that Evelyn had been fined for nonpayment of Society dues, he would have been only one of many.[51] The organization was constantly in financial difficulties, since it was entirely dependent on fees and contributions. The most significant lines consist of a couplet Butler had originally used in place of the lines on the "shilling's pension," but which he deleted from the final version.

> And first found out the building Paul's,
> And paving London with sea-coals.

The juxtaposition of the restoration of St. Paul's and a project for paving London streets points definitely to Evelyn. Pepys mentioned on August 4, 1663, that "Paul's is now going to be repaired in good earnest" and on July 25, 1664, that he had just seen "a printed copy of the King's commission for the repair of Paul's." Evelyn was a member of that commission, together with Christopher Wren and others. On August 27, 1666, there is a lengthy entry in Evelyn's journal of a day he

---

[51] Evelyn was usually scrupulous about his responsibilities to the Society. The lines may imply that in one of his many capacities he was on occasion responsible for fining others.

and Wren and four others spent at the church, with the Bishop of London, the Dean of St. Paul's "and severall expert Workmen," who "went about to survey the generall decays of that antient and venerable Church, & to set downe the particulars in writing, what was fit to be don, with the charge thereof." Work on the restoration seems to have started at once, since, when the Great Fire broke out the following month, Evelyn particularly mentioned its "taking hold of St. Paules-Church, to which the Scaffolds contributed exceedingly."

Both Pepys and Evelyn himself mention a plan in which Evelyn was engaged for paving at least one section of London, not with sea coal, to be sure, though Evelyn was interested in that commodity,[52] but with clinker brick. When he was a young man traveling in Holland in August, 1641, Evelyn had showed particular interest in "that goodly Aquae-duct, or river, so curiously wharfed with Clincar'd, (a kind of White sun-bak'd brick) & of which material the spacious streets on either side are paved." In earlier drafts he had described clinker brick as "a kind of white sun-bak'd brick." Evelyn mentions his later scheme on December 2, 1666, when Sir John Kiviet, former burgomaster of Amsterdam who had been knighted by Charles II, "came downe to examine, whither the soile about the river of Thames would be proper to make Clinkar brick with & to treat with me about some accommodations in order to it." On March 6, 1667, Evelyn proposed to the Lord Chancellor "Monsieur Kiviets undertaking to wharfe the whole river of Thames or Key from the Temple to the Tower (as far as the fire destroied) with brick, without piles, both lasting

---

[52] The term "sea-coal" in this period ordinarily implied the mineral as distinguished from charcoal. On July 11, 1656, Evelyn mentioned coming home by Greenwich Ferry and seeing "Sir Jo: Winters new project of Charring Sea-Coale, to burne out the Sulphure and render it Sweete." He described the process in some detail, adding, "What successe it may have time will discover." Sir John Winter, secretary to Henrietta Maria, after the Restoration obtained a monopoly for the production of coke. I wondered at first whether Butler might have been punning upon *clinker* in the sense in which we use it, but the *OED* indicates the first use of the word in that sense in *Phil. Trans.* for 1769.

and ornamental." Pepys tells us of the outcome of the venture
on September 23, 1668:

At noon comes Mr. Evelyn to me, about some business with
the Office, and there in discourse tells me of his loss, to the
value of £500, which he hath met with, in a late attempt of
making of bricks upon an adventure with others, by which he
presumed to have got a great deal of money: so that I see the
most ingenious men may sometimes be mistaken.[53]

This was not the first time that Evelyn had been included
among satirized virtuosi. In the "Ballad of Gresham College"
he had been honored with four stanzas devoted to his book
Fumifugium, a work devoted to a problem only too familiar to
us today:

> He shewes that 'tis the seacoale smoake
> That allways London doth Inviron,
> Which doth our Lungs and Spiritts choake
> Our hangings spoyle, and rust our Iron.
> Lett none att Fumifuge be scoffing
> Who heare att Church our Sunday's Coughing.

The next speaker among the lunarian observers is again
Robert Boyle, included for a different reason. This time Butler is interested not in inventions or discoveries but in style:

> When one, who for his excellence
> In height'ning words, and shad'wing sense,
> And magnifying all he writ
> With curious microscopic wit,
> Was magnify'd himself no less
> In home and foreign colleges,
> Began, transported with the twang
> Of his own trillo, thus t' harangue.

The reason for my suggested identification is that Butler
(probably a little earlier than this) unquestionably parodied
Boyle's style in his *Reflections* on Dr. Charleton, which I shall

---

[53] Still another sequel to the unfortunate venture would seem to be implied
in the fact that in 1688 Evelyn subscribed 50,000 bricks for the building of a
college for the Royal Society.

consider in another context. At that time I shall return to Butler's strictures on Boyle as a stylist.[54]

Only one other actor in the little drama is described at sufficient length to warrant an attempt at identification. This one
is clearly Robert Hooke for reasons obvious to us, as to Butler's
generation:

> one, whose task was to determine,
> And solve th' appearances of vermin,
> Who'd made profound discoveries
> In frogs, and toads, and rats, and mice,
> (Though not so curious, 'tis true,
> As many a wise rat-catcher knew).[55]

Hooke is one of the few among the speakers who carries over
his supposed "character" into his lunar observations:

[54] It would be possible to make a case for Joseph Glanvill as the speaker
Butler satirized for style. R. F. Jones in his article on "Science and English
Prose Style" (reprinted in *Seventeenth Century Studies in the History of
English Thought and Literature from Bacon to Pope* [Stanford, Calif., 1951],
pp. 75–110) said that Glanvill had been refused membership in the Royal
Society because of his elaborate and ornate style and rewrote his *Vanity of
Dogmatizing (Scepsis Scientifica)* in much plainer style. It is interesting to
notice that Glanvill's style, even in his later period, is mentioned on various
occasions by correspondents of Robert Boyle. On December 10, 1665, Oldenburg
wrote of the "pretty long dedication" made by him to the Society. On October
31, 1666, J. Beal wrote Boyle of a proposal Glanvill had made for another
book and added, "He hath a flowing pen, and may do well if we can ballast
him from Origenian Platonism and extravagant adventure. . . . His genius is
apt for sublime adventure." Oldenburg wrote Boyle again about Glanvill on
October 1, 1667, calling him "a florid writer" (Boyle, *Works*, V, 328 b, 488 a,
367 b). Glanvill can hardly be said to have been famous at home and at foreign
colleges, as the satirist suggests.

[55] The Hooke passage is ll. 377–94 in the octosyllabic version; nothing significant is added in the long-verse. One speaker who, of course, cannot be
identified is the one mentioned in ll. 261–62 merely as "a man of great renown." The other, equally unidentifiable in octosyllabics, is in ll. 291–92,
"another of great worth, / Famed for his learned works put forth." Two lines
added in the long-verse, where the passage occurs ll. 295–99, have caused me
more labor than any other passage in the satire:
> In which the mannerly and modest author
> Quotes the Right Worshipful, his elder brother.
More than one member of the Royal Society was a bishop in his own right,
but I have found no one among the well-known members who was the younger
brother of an archbishop or bishop, though Sir Paul Neile was the son of the
Archbishop of York. I do not find "Sir Poll" in "The Elephant."

> For though the Elephant, as beast,
> Belongs of right to all the rest,
> The mouse, being but a vermin, none
> Has title to it but I alone;
> And therefore hope I may be heard,
> In my own province, with regard.

Into Hooke's mouth Butler also puts words indicating the virtuosi's awareness of the kind of satire directed against them —the kind reported by Pepys that evening when the King laughed at Gresham College:

> It is no wonder we're cry'd down,
> And made the talk of all the Town
> That rants and swears, for all our great
> Attempts, we have done nothing yet.

A little later in the same speech Hooke is made to anticipate some of the satire that may be brought against their present efforts:

> For truth has always danger in't,
> And here, perhaps, may cross some hint
> We have already agreed upon,
> And vainly frustrate all we've done,
> Only to make new work for Stubbes
> And all the academic clubs.

I shall later discuss the last lines, which refer to a savage attack launched against the Royal Society by Henry Stubbe in 1670–71. For the moment I raise the question of the possible date of "The Elephant in the Moon." In its present form the long-verse version could not have been earlier than 1670–71 because of the reference to the Stubbe controversy.[56] But if I correctly interpret Thyer's editorial policy, the reference to Stubbe was not originally in the octosyllabic version. Thyer

[56] The most recent editor of "The Elephant in the Moon" dates the poem (the two versions of which he considers as a whole) not earlier than 1670–71 on the basis of the Stubbe allusion and thinks it may have been as late as 1676. See *Samuel Butler, Three Poems*, selected, with an introduction, by Alexander C. Spence (Augustan Reprint Society. Clark Memorial Library, No. 88; Los Angeles, Calif., 1961).

indicates that he added two italicized words from the long-verse:

> Only to make new Work for *Stubs,*
> And all the *academick* Clubs.

I am myself persuaded that the octosyllabic version was written several years earlier than 1670–71, probably in 1666, not long after Butler had written his prose satire on Dr. Charleton and Boyle's style, which I shall date as 1665.[57] It is significant that all the speakers who can be identified were founding members of the Society, which by 1670 had grown to much greater proportions. Most important to me is the feeling tone of the satire on lunar observation. In its early years the members were much more interested in telescopic observation than during the next decade, when the microscope became more engrossing than the long-familiar telescope. In 1665 and 1666 interest in the moon, planets, comets, and other celestial bodies ran high. In September, 1664, Sir Paul Neile had presented the fifty-foot telescope, through which Hooke and others were for a time busily engaged in checking their earlier observations and adding others. We remember that the comet of 1664 had attracted wide attention, and that another was observed in 1665. The *Philosophical Transactions* for 1665 and 1666 are unusually rich in telescopic papers of a kind that would have caused public interest. Indeed, the first number of the *Transactions,* in 1665, opens with papers on optical instruments and includes an important report on Jupiter and Saturn. Number 2 for 1666 begins with a discussion of "the late comet and a new one." Interest continues well into the following year. A solar eclipse on September 9, 1666, inevitably attracted widespread popular as well as scientific interest, as had a lunar eclipse the

---

[57] The only lines in the octosyllabic version that could not have been written in 1665 are the few on blood transfusion, ll. 213–18, which could readily have been added in 1666 after the dog transfusions took place. It is significant for an early dating, I think, that Butler refers only to dog—not to human—transfusion, as he probably would have had he been writing after 1667.

preceding year.[58] In late 1665 and 1666, I strongly suspect, both the satire on Charleton and one version of "The Elephant in the Moon" were circulating in manuscript among the Wits, a hypothesis which would serve to explain Zachary Grey's later confusion between the tales supposedly told of Sir Paul Neile.

IV

One further satire of Butler's upon scientific enthusiasms of the day—this time in prose—will serve for transition, as I return to the subject of the second essay, infusions and transfusions. I remind you that Pepys had written on March 15, 1665, "Anon to Gresham College, where, among other good discourse, there was tried the great poyson of Maccassa upon a dogg," and also that Dr. Charleton, to whom the poison had been entrusted, had taken it home with him, contrary to all regulations, and had been summoned to appear at the next meeting for discipline. I remind you, too, that as in the instructions Boyle later reported from Lower in connection with the first animal transfusion, operators were instructed "constantly to observe the Pulse . . . in the Dogs Jugular Vein," so infusers like Wren must have watched the pulses of their animals. With these in mind, we may turn to Butler's "An Occasional Reflection on Dr. Charlton's Feeling a Dog's Pulse at Gresham College. By R. B. Esq."

No member of the Royal Society could have failed to recognize what Butler was doing. The title, "An Occasional Reflection" was one Boyle used more than once, particularly in a short paper, published with his "Rhapsody," *Seraphick Love,*

---

[58] Hooke is a better barometer of the interests of the Royal Society than are the *Transactions,* since Oldenburg could publish only a small number of the papers that were read or otherwise presented. Gunther, in *Hooke,* lists approximately thirty-five astronomical entries for 1665–66, many of them extended observations on Mars, Saturn, and Jupiter. Hooke was also closely engaged during those years in improvement of optical instruments. In 1671 Gunther includes only twelve references to any kind of celestial observation, most of them brief and of little importance; in 1672 he includes only eight.

which begins, "My dearest Lindamor." [59] Even the most loyal admirer of Robert Boyle, scientist and philosopher, must have realized how wickedly well Butler parodied his prose style, particularly the style of more personal works such as this. *Seraphick Love,* which Boyle published in 1659, had been written more than a decade earlier, when Boyle was only twenty-one, ambitiously intended as one of a series of moral treatments of sacred and profane love. His prolixity and sentimentality are at their worst in "his theological treatise in romance form," as Marie Boas calls it.[60] The verbosity and moralizing are almost as obvious in the *Occasional Reflections* of 1664, parodied by Swift (some of these, too, had been written early), and appear to some extent in more technical scientific writing. Too often Boyle is obviously "transported with the twang / Of his own trillo," as was his prototype in "The Elephant in the Moon." Thyer said, when he first edited this paper, "His greatest Admirers must confess, that his Stile is rather too copious, diffusive, and circumstantial." Butler, he goes on, "has very archly imitated him both in the flimsy long-winded turn of the Sentences, and in the too pompous Manner of moralizing upon every Occasion that offers."

As we read the "Reflection," we seem to be present with Pepys, Charleton, and Boyle at an infusion performed at Arundel House in the spring of 1665:

TO   Lyndamore

Do you observe, Lyndamore, that domestic animal, the Vassal and menial Servant of Man, on whom he waits like a Lacquey by Day, and watches like a Constable by Night, how quiet and unconcerned it stands whilst the industrious and accurate Dr. Charlton with his judicious Finger examines the arterial Pulsation of its left Foreleg; a civil Office, wherein both Doctor and Dog, Physician and Patient with equal Industry contest, who shall contribute most to the experimental Improvement of this learned and illustrious Society. Little doth

---

[59] "An Occasional Reflection upon a Letter received in April, 1662," is appended to *Seraphick Love,* in *Works,* I, 155–87, 188–90. I have used the Thyer (1759) text of Butler's "Occasional Reflection," I, 405–10.

[60] *Robert Boyle and Seventeenth-Century Chemistry* (Cambridge, 1958), p. 14.

the innocent Creature know, and as little seems to care to know, whether the ingenious Dr. doth it out of a sedulous Regard for his Patient's Health, or his own proper Emolument; 'tis enough to him that he does his Duty; and in that may teach us, to resign ourselves wholly to advance the Interests and Utility of this renowned and royal Assembly.

As the infusion begins, we admire the "Ingenuity of the acute and profound Dr.," who is careful to apply the "Unguent" to a part of the dog's neck nearest the brain, and particularly "furthest out of the Reach of that natural Chirurgery, as I may call it, of his Tongue." Inevitably the satirist, like Boyle, is led to "meditate" upon profound matters: a dog's leg, for example, which in vulgar parlance signifies nothing. Not so to a philosopher: to him that homely, familiar object may teach man that there is nothing contemptible in Nature, that all things may "contribute something to the public Good of Mankind, and Commonwealth of Learning." He meditates upon dogs *versus* cats: how wise the doctor was in not choosing as his subject "that vigorous and vivid Animal commonly called a Cat." A cat, after all, has nine lives. How could the operator hope that the "lethal Force of this destructive Medicament" would kill all nine? Perhaps, indeed, this was the explanation for the comparative failure of an infusion into a kitten at another meeting of the Royal Society which you and I also attended with Pepys, when the "Florence poyson" was administered to a hen, a dog, and a cat. Pepys left before the end of the experiment, but Butler's informant had remained throughout, and his report that the kitten became drowsy gave the satirist opportunity for a climax the secretary had not mentioned in the official minutes:

when we last tryed this very Experiment on a Creature of that Species, although but a weak and feeble Kitten, the venomous Quality proved so innocuous, that the secure little Beast laid it self down to sleep in the hollow Concave of that Emblem of our Jurisdiction over the Lives and Limbs of Dogs and Cats, the Mace; and in that Posture, as if it had triumphed over its mortal Enemy, and all our Hostilities, was borne before the most excellent and accomplished Lord President.

Butler sees to it that Charleton is aware that the disciplinary action of the Society toward his infraction of rules was known outside the walls of Arundel House. So impatient was the doctor that he could not wait the date set but took the poison home, presumably to "try the Experiment solitary." "For which he received, I will not say whether condign Punishment, or severe Castigation, in a grave and weighty Oration pronounced by his Lordship before this celebrious and renowned Assembly."

In his penultimate paragraph the satirist unkindly reminds the virtuosi of one of many comments made by the French visitor, Samuel Sorbière, whose *Relation* of what he had seen at the Royal Society had caused resentment on the part of various members of the Society. Sorbière had considered it "a Work of Admiration" that a gentleman of title, one "bred up in Courts and Camps, and at this present employed in the most weighty Affairs of State," should be seen setting up Society telescopes in St. James' Park, "to Muster the Life-guard of Jupiter, and to take an Account of the Spots in his Belt." It seems no less remarkable that this "exquisite and solert Dr. whose Province lies in the Cabinet of fair Ladies" and whose usual employments are "to sollicit the tender Arteries of their Ivory Wrists" should, for the sake of science, demean himself and "condescend to animadvert the languishing Diastole of an expiring Mungrel."

In conclusion "R. B." meditates profoundly upon the fact that in Nature there is presumably neither height nor depth nor any other creature, but that all things, great and small, contribute to the knowledge of virtuosi. Their father, Francis Bacon, he might have added, had warned his disciples not to despise "mean and even filthy things," a lesson followed by his sons, particularly by one:

the most industrious and elegant Mr. Hook, in his Microscopical Observations, has most ingeniously and wittily made it appear, that there is no difference, in point of Design and Project, between the most ambitious and aspiring Politician of the World, and of our Times especially, and that most importune and vexatious Insect, commonly called a Louse.

"An Occasional Reflection" is one of the few of Butler's minor works that can be dated with a fair degree of accuracy. The satire is too "occasional" to have been written much later than the reported events occurred. The Macassar poison had been tried on March 15, 1665, and Charleton's discipline took place the following week. The comparative failure of the Florentine poison upon a kitten occurred the next month, on April 19. Early in the same year Hooke's *Micrographia* had appeared, with its extraordinary enlargements of fleas, lice, and other vermin. There is little doubt that in 1665, while the stories about Charleton, the dog, and the kitten were current, wits in the coffeehouses were snickering over the style of "R. B." to his "dearest Lyndamore." [61]

V

It was, of course, inevitable that the Royal Society experiments on blood transfusion should have been a talk of the town. Butler had introduced lines on dog transfusion into "Hudibras to Sidrophel":

> Can no transfusion of the blood,
> That makes fools cattle, do you good?
> Not putting pigs to a bitch to nurse,
> To turn them into mongrel curs? . . .
> As if the art you have so long
> Profess'd, of making old dogs young,
> If you had virtue to renew
> Not only youth, but childhood too.[62]

[61] I have not included any study of Part III of *Hudibras*, in part because it lies beyond the chronological scope of this book, since it was not published until 1678, and in part because little scientific satire appears.

[62] Lines 39–62. The couplet about "putting pigs to bitch to nurse" could have been written at any time, since the belief that the milk of nurses affected their charges was very old. We remember Pepys' repeating the story about Dr. Caius. The other lines were not earlier than November 1666, when the

In "The Elephant in the Moon," Butler, as so often, repeated some of the phrases together with a variant on others:

> No more our making old dogs young
> Make men suspect us still i' th' wrong . . .
> Nor putting dogs t' a bitch to nurse,
> To turn them into mongrel-curs.

It is, I think, significant that this passage was entirely deleted from the long-verse version of "The Elephant" and its place taken by references to two other Royal Society experiments which had nothing to do with transfusion.[63] I suspect this fact offers a further clue to dating the long-verse version not earlier than 1670–71, and possibly later. Transfusion had been legally banned in France in 1670. Whether or not it was actually forbidden in England, shortly after the French furore references to it cease in the minutes of the Royal Society.[64]

The Wits were not the only adversaries whose opposition the virtuosi rightly feared. They were set upon by various members of the univerities, who considered that the New Philosophy was attempting to undermine the True Philosophy, long taught within their walls, where traditionalists opposed the experimental method as long as possible. The early "Ballad of Gresham College" should have afforded them some ammunition:

---

first transfusions were performed, and probably a little later, after there had been talk about making "old dogs young."

[63] Lines 213 ff. in the octosyllabic version have been amended in the long-verse—ll. 211 ff.—by the excision of the references to transfusion and the addition of

> Nor little Stories gain Belief among
> Our criticalest Judges right or wrong.

In ll. 217–20 Butler replaced the lines on transfusion by

> Make Chips of Elms produce the largest Trees,
> Or sowing Saw-dust furnish Nurseries.

Thyer, *Genuine Remains*, I, 38 n., points out correctly that this is exaggerated from an account of "Elms Growing from Chips," included in Sprat's *History*, p. 197.

[64] Following this, in ll. 219–24, Butler included another passage based on two tales inserted in Sprat's *History*, pp. 212–13, "A Relation of the Pico Teneriffe."

> Thy Colledg, Gresham, shall hereafter
> Be the whole world's Universitie,
> Oxford and Cambridge are but laughter:
> Their learning is but Pedantry.

A half-dozen years later John Evelyn wrote at length in his journal about the dedication of the Sheldonian Theatre at Oxford, one of Christopher Wren's first academic buildings, on July 9, 1669, an occasion of great splendor and formality, that drew "a world of strangers" to the University. The University Orator, Dr. Robert South, delivered an overlong address, "not without some malicious & undecent reflections on the Royal Society as underminers of the University, which was very foolish and untrue, as well as unseasonable." [65] John Wallis also described the exercises in a letter to Robert Boyle on July 14, 1669, saying that

Dr. South, as university orator, made a long oration. The first part of which consisted of satyrical invectives against Cromwell, fanaticks, the Royal Society, and new philosophy; the next of encomiasticks, in praise of the archbishop, the theatre, the vice-chancellor, the architect, and the painter; the last, of execrations against fanaticks, conventicles, comprehension, and new philosophy; damning them *ad inferos, ad gehennam.*

The Royal Society was opposed, too, by members of the Royal College of Physicians—not by all, by any manner of means, since there was decided overlapping in the membership of the two organizations. In England, as in Paris, many physicians, like their counterparts in the universities, were hidebound traditionalists. Even some of the more liberal believed that medical and anatomical matters should be in the hands of the College of Physicians and that the virtuosi were trespassing

[65] The only sentence known from this oration is one Disraeli quoted: "[The Virtuosi] can admire nothing except fleas, lice and themselves." Apparently the speech was extant in the nineteenth century, though it seems to have disappeared. Dorothy Stimson, who quotes the Disraeli line, says (*Scientists and Amateurs,* p. 77): "It is a matter of regret that prolonged search in 1930 failed to unearth this oration." The Wallis account is in Boyle, *Works,* V, 514–15.

on their terrain, sometimes dangerously, as in the experiments on transfusion. The right-wing position is well illustrated by a passage in a biography of Dr. Baldwin Hamey, a member and benefactor of the College of Physicians:

> It griev'd [Hamey] to foresee a Rival Society treading close upon the heels of the Aesculapians, whose vortex would be so great as to comprehend everything, as indeed it came to pass; whereas all matters within the sphaere of Medicine, Anatomy and Surgery, most properly should belong to the Royal College of Physicians, whilst the largest purlieus of Natural Philosophy and Mathematicks, exclusive of the former, might have found the Societists Employment and Enquiry enough.[66]

The opposition the Society incurred from some physicians—particularly those inclined to exorbitant quackery in their prescriptions—and even more from apothecaries was the result of belief that the Society members might follow their ancestors of Bacon's *New Atlantis* too far in their enthusiasm for synthetic medicines, or their own "quackeries," as Butler implied in the "Satire on the Royal society":

> To make a powder of the sun
> By which all doctors should b' undone.

We remember the conversation Pepys reported between a group of doctors and a group of apothecaries and recall that not only were the apothecaries even more rigidly antiquated than the doctors but also that they proved more vocal.

With a desire to avert or reply to such sources of criticism and satire the Royal Society, about 1664, had commissioned Thomas Sprat to write its *History*. Although the book was well received by the *avant garde*, it was far from putting an end to either criticism or satire, which, indeed, it provoked. One of the most vocal adversaries of both the Society and the *History* was Robert Crosse, vicar of Great Chew in Somersetshire, a confirmed Aristotelian, in every way an "Ancient" of the

---

[66] Quoted in Harcourt Brown, *Scientific Organizations in Seventeenth Century France* (Baltimore, 1934), pp. 255–56; Cope, *Joseph Glanvill*, p. 26.

crusted tawny-port variety. He was a scholar of distinction who had refused the Regius Professorship of Divinity at Oxford to settle in his small parish. Crosse, we are told,

travelled up and down to tell his Stories of the Royal Society, and to vent his spite against that Honourable Assembly. He took care to inform every Tapster of the Danger of their Designs; and would scarce take his Horse out of an Hostler's hands, till he had first let him know how he had confuted the Virtuosi. He set his everlasting tongue at work in every Coffee-House, and drew the Apron-men about him, as Ballad-Singers do the Rout in Fairs and Markets; They admir'd the man, and wondered what the strange thing call'd the Royal Society could be.[67]

Not far from Robert Crosse lived Joseph Glanvill,[68] rector of the Abbey Church in Bath, who had been admitted to the Royal Society in 1664, a membership of which he was very proud. He had been introduced to Crosse in 1667, a meeting that resulted in mutual acrimonious dislike. Crosse is said to have written a book, with the virtuosi and Glanvill as covillains of the piece, a book said to be so scurrilous and vituperative that it was refused a license.[69] In 1668 Glanvill published *Plus Ultra; or, The Progress and Advancement of Knowledge*

[67] Joseph Glanvill, *A Praefatory Answer to Mr. Henry Stubbe* (London, 1671), p. 3.

[68] The relationship between Glanvill and Pepys is somewhat mysterious. Glanvill's name appears in the *Diary* at least ten times during the autumn of 1665. On various occasions Pepys mentions spending an evening or the night in Glanvill's house—then in London. At no time does Glanvill himself seem to have been present. His house was used by Pepys and Captain George Cock as a repository for their share in the distribution of Dutch goods brought by Lord Sandwich's fleet. (See Arthur Bryant, *Samuel Pepys: The Man in the Making* [New York, 1933], pp. 264–72.) The association with Glanvill seems to have been on Cock's part rather than on Pepys'. Later in the *Diary* Pepys mentions some of Glanvill's writings, particularly on the invisible drummer and witches. Pepys and Glanvill must have had some association through their membership in the Royal Society.

[69] Glanvill says in *Plus Ultra* that he himself sent the contents of Crosse's book to a friend in London, who published fewer than one hundred copies as the *Chew Gazette*. Wood, *Athenae Oxonienses*, pp. 569–70, discusses the Crosse-Glanvill-Stubbe controversy, and mentions the publication of the "Chue Gazette." No copy of the gazette seems to have survived, since Professors Jones and Cope both indicate that they were unable to find it.

since the Days of Aristotle. Although he indicated that the work had been "occasioned by a Conference with one of the Notional Way"—obviously Robert Crosse—there is evidence that Glanvill had intended such a work before Sprat's *History* was published. Oldenburg mentioned the proposal in a letter to Boyle, October 1, 1667, saying that the book would have "been extant, I find by his letter, ere this; but that he stayed for Mr. Sprat, to see what room he would leave for his thoughts, and finding now, that he had not throughout prevented him, he seems resolved to pursue his design." [70]

One of Glanvill's basic points of departure for his defense of the Royal Society was the charge the King had brought in the Duke's chamber, which we have heard also in Butler. Glanvill raises the problem: "an insulting Objection that we hear frequently in this Question, What have they done?" But Glanvill answers the question. Indeed, the whole volume may be said to be an elaboration and development of one succinct sentence here: "To this I could answer in short . . . more than all the Philosophers of the Notional way, since Aristotle opened his Shop in Greece." [71] Glanvill's *Plus Ultra* was well received by the virtuosi. Evelyn wrote on June 24, 1668,[72] thanking him for a copy Glanvill had sent him and praising

this worthy vindication both of yourself and all useful learning against the science (falsely so called) of your snarling adversary. I do not conceive why the Royal Society should any more concern themselves for the empty and malicious cavils of these delators, after what you have said; but let the moon-dogs bark on, till their throats are dry: the Society every day emerges.

[70] Boyle, *Works*, V, 367.

[71] *Plus Ultra*, p. 90. In connection with Glanvill's obvious enthusiasm for the accomplishments of the Royal Society, it is interesting to notice the mood of Oldenburg's earlier letter to Boyle about Glanvill's dedication of his *Scepsis Scientifica* to the Society: "In which dedication the author expresseth a very great respect to the said body and their design; which I was very glad, and so were others, to find themselves to be so well understood, at last, by some, though I fear the great expectation he raiseth of their enterprise, may be of more prejudice than advantage to them, if they be not competently endowed with a revenue to carry on their undertakings" (Boyle, *Works*, V, 328 b).

[72] *Diary and Correspondence of John Evelyn*, ed. Bray (London, 1875), III, 204. Quoted in Cope, p. 23.

From the time he published *Plus Ultra* Glanvill was pursued by Crosse as relentlessly as any hare by hounds. Failing to find a publisher for his book, says Wood,[73]

Crosse wrot Ballads against him, and made him and his Society ridiculous; while other Wags at Oxon, who seemed to be pleased with these Controversies, made a dogrel Ballad on them and their proceedings; the beginning of which is,
>   Two Gospel Knights
>   Both learned wights
>   And Somerset's renowne a,
>   The one in Village of the Shire
>   But Vicaridge too great I fear,
>   The other lives in towne a.

Soon Glanvill himself declared that "there was no other subject handled on Ale-benches, and in Coffee-Houses" except Glanvill's supposed atheism.[74] His position in his parish became increasingly difficult. Then entered the arena a bitter enemy of Glanvill and a staunch supporter of Crosse, vocal in vituperation, so rapid in writing that he did indeed pour himself forth with that facility that should be stopped. In 1670–71 spewed from the press no fewer than seven books and pamphlets, several of them personal attacks on Glanvill, all directed against the Royal Society.[75] Henry Stubbe (Stubb, Stubbs,

---

[73] Page 570.

[74] Quoted Cope, p. 24. I cannot agree with Cope's attribution (p. 36) to Glanvill of a ballad, "The Character of a Coffee-House," as I cannot agree with him about Glanvill's authorship of the "Ballad of Gresham College." Cope dates the verses as of 1673. Gwendolen Murphy, *A Cabinet of Characters* (London, 1925), pp. 315–22, gives the first version, "News from the Coffee-House" (London, 1667), with others in 1672 and 1673. Samuel Weiss, *Notes and Queries*, CXCVII (1952), 234–35, 343, pointed out verbal similarities between the ballad and some of Glanvill's work and concluded that either Glanvill was the author or that the anonymous versifier plagiarized from him, which seems much more probable. In 1667, attacked as Glanvill was in coffeehouses, it seems incredible that he should have written what is largely a lighthearted satire. The ballad also has been published in *Harleian Miscellany* (London, 1810), VIII, 7–13.

[75] According to the account given by the nephew of Dr. Baldwin Hamey, from which I have quoted, Hamey was responsible for Stubbe's entering the controversy. Cope quotes, p. 26: "Dr. Hamey therefore found out a person of his own Profession but a Country Practiser, one Dr. Henry Stubbs, a Man of as much Acrimony as Wit, with as knowing a head, as he had an able hand

Stubbes) was a physician in Warwick, who had a summer practice in Bath. During part of the year, at least, the three antagonists were living within a few miles of each other. In addition to being a doctor, Stubbe was also an excellent classical scholar and had formerly been sublibrarian of the Bodlein. He was, said Anthony à Wood,

a very bold man, utter'd any thing that came into his mind, not only among his Companions, but in publick Coffey-houses, (of which he was a great frequenter). . . . He had a hot and restless head (his hair being carrot-colour'd) and was ever ready to undergo any enterprize, which was the chief reason that macerated his body almost to a Skeleton. He was also a person of no fix'd Principles.[76]

Here, then, is the cast of chracters, as well as the setting and the incipient plot of the melodrama [77] to which Butler was referring when "Hooke" said:

> Only to make new work for Stubbes,
> And all the academic clubs.

Since the "Stubbe controversy" has been discussed in detail by highly competent scholars, I shall make no attempt to enter into it, except for one aspect which Professors Jones, Baker, and Cope mention, when at all, merely in passing: Stubbe's pronounced opposition to the Society on medical subjects. He was, after all, a doctor, and evidently a good one, since, when most members of the Society were antagonistic to him, he and Robert Boyle remained on good terms, Boyle respecting him

---

. . . and this man he generously retayn'd for his Champion against the Royal Society: Stubbs then drew his Pen with great virulence, and lay'd it about him most furiously indeed, and was well gratified by Dr. Hamey for it." The implication that Hamey paid Stubbe for his labors may have inspired the statement of Brown, *Scientific Organizations*, pp. 256–57, that Stubbe was hired to attack the Royal Society. Whether or not money played a part, it was a minor one, since there is no question of the sincerity of Stubbe's attacks on the Society.

[76] Page 415.

[77] The Stubbe controversy was first discussed by R. F. Jones in the 1936 edition of *Ancients and Moderns*. It has been thoroughly treated by Cope in *Joseph Glanvill* and discussed also by Herschel Baker, *The Wars of Truth* (Cambridge, Mass., 1952), pp. 535–66. There is a complete bibliography of the pamphlets and books in Cope, p. 27, n. 82.

as a doctor whose patient he had been on occasion. Other members of the Royal Society, who devoted their energies to astronomy, physics, and chemistry, might find serious gaps in Stubbe's knowledge, but in anatomy and medicine he was on safe ground, more secure than was Joseph Glanvill.

In a postcript appended to *Campanella Revived*,[78] Stubbe entered into the opposition that the Royal Society was encountering both from members of the Royal College of Physicians and from apothecaries. He believed with Dr. Hamey that the Society was encroaching upon the domain of the College, to the discredit of physicians and possible danger to patients. He attacked the "errours and cheat of the Virtuosi . . . the tendencies of whose design was so fatal and malignant." "The blind," he wrote, "may as well judge of colours, the insensible concerning the objects of feeling, as the Virtuosi of Physick." To some extent in this pamphlet, but much more in his long answer to Glanvill's *Plus Ultra*,[79] in which Glanvill had praised the Royal Society as the pioneer and innovator in infusion and transfusion, Stubbe entered into detailed discussion of these recent innovations in medicine, showing a great deal of knowledge of both processes.

Stubbe's point of departure was a statement made by Glanvill that the Royal Society had inaugurated both infusion and transfusion. To some extent, of course, Stubbe was quibbling, yet his insistence that neither process was original with the Society we know to be correct. It is interesting to find that

---

[78] *Campanella Revived; or, An Enquiry into the History of the Royal Society* (London, 1670). The postscript deals particulary with a "Quarrel Depending betwixt H. S. and Dr. Merrett," pp. 19–22. Merrett, who had been elected to the Royal Society in 1663, had evidently been opposing the apothecaries, who believed that the College of Physicians was deliberately trying to abolish their profession, preparing their own prescriptions, some for caution, some for financial gain.

[79] *The "Plus Ultra" Reduced to a Non Plus; or, A Specimen of Some Animadversions upon the "Plus Ultra" of Mr. Glanvill, wherein Sundry Errors of Some Virtuosi Are Discovered . . .* (London, 1670). The two parts have separate title pages, but are bound together. There is an extensive section on infusion and transfusion, pp. 116–79, concluding with the letter signed "Agnus Coga."

Stubbe spoke from firsthand knowledge so far as infusion was concerned. This, he says, was 'a thing much practised by Dr. Wren and others in Oxford, before the Restoration of his Majesty, and before that ever the Society was thought upon." This fact was "well known to all that were at those days in the University," as Stubbe himself had been, since he matriculated at Christ College in March 1650/51 and proceeded B.A. in 1653. He had also returned to Oxford as assistant to the librarian of the Bodleian in 1657–59. He does not indicate during which of these periods he himself had seen "the Dog into whose veins there was injected a Solution of Opium, at the Lodgings of the Honourable Robert Boyle."

So far as transfusion was concerned, in his zeal to strip off all claim to originality from the Royal Society, Stubbe, like many others of the period, insists that priority for that process belonged to Libavius, who had described the process "above Fifty years ago." While Libavius described transfusion, he did not perform it. Priority in this respect, as we have seen, belonged to Richard Lower. Stubbe had no intention of detracting from Lower's importance in the field. Indeed, he makes clear that he greatly admires Lower, with whom he had personal acquaintance.[80] He mentions the date on which a possible experiment in transfusion was first mentioned by the Society and is entirely familiar with the fact that Lower and Boyle had instructed the doctors who performed the Royal Society transfusion. Much of the long section on infusion and transfusion was devoted by Stubbe to discussion of the possible advantages involved in the processes, which are, he insists, far outweighed by the dangers to human health and life. He discusses in detail what we call "incompatibility" of bloods: animal blood is not homogeneous with that of human beings, and the blood of some animals is heterogeneous with that of others. Stubbe knew whereof he spoke, since he had evidently spent a

[80] At the end of his discussion of infusion and transfusion, on p. 178, after having indicated that Willis had had little to do with these processes, he concludes: "Thus much I thought fitting to annex, lest the Virtuosi should censure me as partial to my old School-fellow, Dr. Lower."

great deal of time at one period of his life experimenting with the effect of various infusions upon animals, accounts of some of which he sets down in detail. At one point he indicates that as early as 1660 he had gone farther than the Royal Society ever went with such experiments. While he does not claim priority in transfusion, there is no question that he was greatly interested in blood analysis and had observed blood carefully in connection with his patients. Quibbling aside, Stubbe comes off much the victor over Glanvill in this section on anatomy and medicine, since he speaks with authority both as a practicing physician and as one who has done original research in the field. His general conclusion is that the dangers inherent in both infusion and transfusion far outweigh any advantage to be gained by their trial:

There was a time when men had regard to their Consciences, and what could not be administered but upon prudential hopes of advantage to the Patient, no approved Physician durst, or would give to any sick person: but in this Age, such as ought to protest against it, are as forward as any to forget these considerations, and prompt men on to practises without either regarding whether the effect be not Murther in the Physicians, besides the ill consequences to be diseased. In the injection of Medicaments, I must complain that neither the Operation of Medicaments immediately injected into the blood and veins is known, nor the dose; and consequently the Project not like to improve Physick at all, unless our Magistrates will licence men to try so many Experiments, even to the apparent hazard or certain death of the parties.[81]

Stubbe is responsible for our hearing the final chapter in the story of Arthur Coga and the blood transfusions. The laughter

---

[81] *Ibid.,* p. 122. In *Campenella Revived,* Preface, Stubbe took exception to the pride of the Royal Society in such respiratory experiments as those of Hooke and the bellows and Boyle and the broken-winded dog, although he mentions specifically experiments by "Dr. Croune and Dr. Thruston." The "experiment about reviving a strangled Fowl by blowing ayre into the lungs" was as old as Vesalius, he asserts, and "vulgarly known before, though perhaps not to the Virtuosi." As a practicing physician, he adds that the principle was also well known to and used by midwives: "when Children are Still-born, or any one blow into their Breast."

of Butler in his little fragments would have been only irritating to the Royal Society. There is evidence, however, that the Wits had had a serious effect upon their medical plans in connection with Coga. Lower, you may recall, said that, in the hope of improving Coga's mental condition, he had intended to repeat the treatment several times. "He, on the other hand, consulted his instinct rather than the interests of his health, and completely eluded our expectations." We find the reason Coga did not return for later transfusions in the correspondence of another distinguished scientist, John Ray, the greatest naturalist of the century. On December 13, 1667, a former pupil of his, Sir John Skippon, wrote to him of a meeting of the Royal Society, to which he had recently been elected:

The dean [John Wilkins] says he is confident no man can translate his book, "Real Character," better than yourself. Yesterday the transfusion of blood was experimented upon the same body they hired at first: they let out eight or ten ounces of his own, and then transfused of the sheep's arterial blood about fourteen or sixteen ounces. There was a great company present.[82]

In a later letter to Ray, written apparently in the following January, Skippon, who had attended another meeting of the Royal Society, wrote: "The effects of the transfusion are not seen, the coffee-houses having endeavoured to debauch the fellow, and so consequently discredit the Royal Society and make the experiment ridiculous." Boyle had spoken only too truly when he said that some Wits used the Muses like anchovies, to tempt men to drink. It was, I suspect, at a table of the Wits—possibly at their dictation—that Arthur Coga wrote a letter published by Stubbe,[83] who may well have picked up a copy at one of the coffeehouses, "of which," said Anthony à Wood, "he was a great frequenter."

[82] *The Correspondence of John Ray*, ed. Edwin Lankester (London, 1848), p. 22. The second letter of Skippon to Ray is not dated, but is immediately above another dated "Jan. 24, 1668."

[83] *A Specimen of Some Animadversions upon a Book, Entitled "Plus Ultra,"* . . . (London, 1670), p. 179.

To the Royal Society of Virtuosi, and all the
Honourable Members of it, the Humble Address of
Agnus Coga.

Your Creature (for he was his own man till your Experiment
transform'd him into another species) amongst those many alterations he finds in his condition, which he thinks himself
oblig'd to represent them, finds a decay in his purse as well as
his body, and to recruit his spirits is forc'd to forfeit his nerves,
for so is money as well in peace as warre. 'Tis very miserable,
that the want of natural heat should rob him of his artificial
too: But such is his case; to repair his own ruines, (yours, because made by you) he pawns his cloaths, and dearly purchases
your sheeps blood with the loss of his own wooll. In this ship-
wrack't vessel of his, like that of Argos, he addresses himself to
you for the Golden Fleece. For he thinks it requisite to your
Honours, as perfect Metaplasts, to transform him without as
well as within. If you oblige him in this, he hath more blood
still at your service, provided it may be his own, that it may be
the nobler sacrifice.

The meanest of your Flock,
Agnus Coga.

And so "Lamb" Coga disappears, not upon an operating table
in Arundel House, where he had been expected, but probably
under the table in a tavern or coffeehouse.

VI

There are many important reasons for regret that Pepys could
not continue his *Diary* after May 30, 1669; one of my own personal disappointments is that I shall never know his reaction
to a play he must have attended, Thomas Shadwell's *The Virtuoso*,[84] first performed in May, 1676, by the Duke's Company
at the Dorset Garden Theatre. Appropriately enough, it was
dedicated to the Duke of Newcastle. Here is the most extensive
satire on science in Restoration drama, and one of the most ex-

[84] With David Rodes of Stanford University, I recently completed a new
edition of *The Virtuoso*, which will appear as the fifth volume in the Regents
Restoration Drama Series, being published at the University of Nebraska Press.
This is based upon the first edition of 1676, a copy of which is at the Hunting-
ton Library. I have followed here the text as modernized by Rodes.

tensive in any form of popular literature. Inveterate playgoer as he was, Pepys must have been torn between amusement at the farce and loyalty to the Royal Society. The King and the aristocrats, who had laughed privately at Gresham College a dozen years earlier, now had reason to laugh publicly at the scientific inventions, discoveries, and theories of Sir Nicholas Gimcrack, the "virtuoso," in whose character Shadwell wickedly portrayed generic enthusiasms of members of the Royal Society.

The Virtuoso is a typical Restoration comedy, with love plots, concealment scenes, and mistaken identity. I shall deal only with the scientific satire, which is largely in scenes in which Sir Nicholas Gimcrack appears, sometimes with Bruce and Longvil, young lovers who have gained access to Gimcrack's nieces by feigning profound interest in science. Their audience with the master is delayed because Gimcrack is learning to swim, not in the water, but upon a laboratory table, where we first see him, surrounded by "his instruments and fine knacks," lying upon the table, imitating the motions made by a frog in a bowl of water. Asked by Longvil whether he intends to swim in water, he replies, "Never, sir. I hate the water." Then what, inquires his amazed auditor, will be the use of swimming? "I content myself," Gimcrack replies, "with the speculative part of swimming; I care not for the practical. I seldom bring anything to use; 'tis not my way. Knowledge is my ultimate end" (II.ii.86–91). And so, indeed, we discover, as we follow his inventions and discoveries, most of which sound very familiar to us.

Shadwell knew Wilkins' *A World in the Moon.* His Gimcrack not only believed that the moon was an earth, but was about to publish a book of geography about it, proving that it had mountains and valleys, seas and lakes. Shadwell seems also to have read Butler's "Elephant in the Moon," since Gimcrack's telescope had shown lunar elephants and camels and a land battle fought with "Elephants and Castles." Before Sir Nicholas had learned to swim, he had learned to fly, since "a man by art may appropriate any element to himself." He added, quoting from Joseph Glanvill's *Scepsis Scientifica:* "Nay, I doubt not but in a little time to improve the art so far, 'twill be as common to buy a pair of wings to fly to the world in the moon as to buy a pair of wax boots to ride into Sussex with" (II.ii.29–38).

Like Robert Boyle, Gimcrack knew all about the nature of air, which he too had weighed and which he was aware is "but a thinner form of liquor." Like Boyle, too, he had "factors" who traveled widely, collecting samples of air, which they hermetically sealed in bottles, kept by Sir Nicholas in his cellar in place of wine. The Virtuoso is able to enjoy all the advantages of a holiday in the country without the inconveniences of travel: "Now if I have a mind to take country air, I send for maybe forty gallons of Bury air, shut all my windows and doors close, and let it fly in my chamber" (IV.iii.287–90). With the King and the satirists before him, Longvil asks the inevitable question: "But to what end do you weigh this air, sir?" This time he learns the answer: "To what end should I— to know what it weighs. O knowledge is a fine thing" (V.ii. 22–24).

Boyle is parodied, too, in passages over which I shall not stop at length, since Pepys did not happen to give us the background for Boyle's many experiments in luminescence. Any member of the Royal Society would have understood what Sir Nicholas was talking of when he said:

There was a lucid sirloin of beef in the Strand. Foolish people thought it burnt when it only became lucid and crystalline by the coagulation of the aqueous juice of the beef by the corruption that invaded it. 'Tis frequent. I myself have read a Geneva Bible by a leg of pork. . . . I could eclipse the leg of pork in my receiver by pumping out the air. But immediately upon the appulse of air let in again, it becomes lucid as before. [V. ii. 32–43]

Hooke's microscopes and his *Micrographia* come in for a large share of attention. Gimcrack had invented, developed, or purchased as many instruments as Hooke used: microscopes, telescopes, thermometers, barometers, pneumatic engines, stentrophonical tubes. Indeed, he had outdone Hooke in his invention of loud-speakers so powerful that in a short time, he prophesied, one man on a hill might be heard throughout a whole country. This would prove a real service to His Majesty, since when the Gimcrack stentrophonical tube was perfected, only one parson would be needed to preach to a county and the King could take back the church lands and benefices which he was now forced to bestow on chaplains.

Of all instruments, microscopes were Gimcrack's greatest enthusiasm. His resentful nieces, indeed, tell us that their uncle is a sot who has spent two thousand pounds on microscopes to find out the nature of eels in vinegar, mites in a cheese, and the blue of plums, which he has discovered to be living creatures. He has broken his brains over maggots and made a profound study of spiders, but never cares for understanding mankind. Gimcrack's toady boasts: "No man upon the face of the earth is so well seen in the nature of ants, flies, humble-bees, earwigs, millepedes, hog's lice, maggots, mites in a cheese, tadpoles, worms, newts, spiders, and all the noble products of the sun by equivocal generation." Sir Nicholas offers his visitors a lecture on ants, whose eggs he has dissected on the object plate of his microscope. Again we hear the reiterated refrain:

Bruce. What does it concern a man to know the nature of an ant?
Long. O it concerns a virtuoso mightily; so it be knowledge, 'tis no matter of what. [III.iii.1–6, 27–29]

Hooke's scientific language, with that of his colleagues, comes in for parody, particularly in a section from the *Micrographia*. The Virtuoso was studying micrographically the transmutations of a plum, which "comes first to fluidity, then to orbiculation, then fixation, so to angulization, then crystallization, from hence to germination or ebullition, then vegetation, then plantanimation, perfect animation, sensation, local motion, and the like" (IV.iii.243–48). Sir Nicholas rolled the syllables over on his tongue with relish; Shadwell, his creator, changed only one word of Hooke's original, using "local motion" for Hooke's "imagination." [85]

Of the many articles Shadwell must have consulted in the *Philosophical Transactions* and elsewhere, you and I will not be surprised to learn that the ones that interested him most seem to have been those on blood transfusion. Sir Nicholas, like various members of the Royal Society, had begun his studies in transfusion by experiments in respiration. He had "found out the use of respiration or breathing; which is a motion of the thorax and the lungs whereby the air is impell'd by the nose, mouth, and windpipe into the lungs and thence expell'd farther to elaborate the blood by refrigerating it and separating its fuliginous steams." He had presumably anticipated Hooke in the discovery "that an animal may be preserv'd without respiration when the windpipe's cut in two, by follicular impulsion of air; to wit, by blowing wind with a pair of bellows into the lungs" (II.ii.101–11). From respiration Gimcrack had gone on to blood transfusion, beginning with transfusions between animals. Sir Formal, his toady, interrupts to tell the visitors that he has seen Gimcrack "do the most admirable effects in the world upon two animals: the one a domestic animal commonly call'd a mangy spaniel, and a less famelic creature commonly call'd a sound bulldog." In words lifted almost bodily from the *Philosophical Transactions* Shadwell only slightly exaggerates the Lower operation performed by Drs. King and Coxe:

Why I made, sir, both the animals to be emittent and recipient at the same time. After I had made ligatures as hard as I could

[85] This was pointed out by Claude Lloyd, "Shadwell and the Virtuosi," *PMLA*, XLIV (1929), 472–94. Lloyd analyzed many passages in the play with reference to experiments reported in *Phil. Trans.* and elsewhere. A little of the scientific satire had earlier been interpreted by Carson S. Duncan, *The New Science and English Literature in the Classical Period* (Menasha, Wis., 1913). Since Duncan and Lloyd wrote, much more information has been made available by Gunther and others.

(for fear of strangling the animals) to render the jugular veins turgid, I open'd the carotid arteries and jugular veins of both at once time, and so caus'd them to change blood one with another.

Sir Formal joins in the chorus of acclamation: "Indeed that which ensu'd upon the operation was miraculous, for the mangy spaniel became sound and the bulldog mangy" (II.-ii.125–34).

Shadwell had read with profit some of the accounts of Denis's operations in Paris, with the result that, when he came to human transfusions, Gimcrack could speak learnedly of the condition the French physicians discovered in the Swedish Baron Bond: "Those men suffer'd not under the operation, but they were cacochymious and had deprav'd viscera, that is to say, their bowels were gangren'd" (II.ii.232–34). In the climactic performance Shadwell and Sir Nicholas had had the assistance of both Denis in France and Lower in England. When Longvil remarks that "that was a rare experiment of transfusing the blood of a sheep into a mad-man," Sir Nicholas boasts:

Short of many of mine. I assure you I have transfus'd into a human vein 64 ounces, avoirdupois weight, from one sheep. The emittent sheep died under the operation, but the recipient madman is still alive. He suffer'd some disorder at first, the sheep's blood being heterogeneous, but in a short time it became homogeneous with his own. . . . The patient from being maniacal or raging mad became wholly ovine or sheepish: he bleated perpetually and chew'd the cud; he had wool growing on him in great quantities; and a Northamptonshire sheep's tail did soon emerge or arise from his anus or human fundament. [II.ii.196–202, 205–9]

It would seem that Gimcrack's creator had read the letter from "Agnus" Coga, published by Henry Stubbe, since in his climax he mentions one very similar in diction: "Here is a letter from the patient who calls himself the meanest of my flock, and sent me some of his own wool. I shall shortly have a flock of 'em. I'll make all my clothes of 'em; 'tis finer than beaver. Here was one to thank me for the cure by sheep's blood just now" (II.ii.223–28). At this, Gimcrack's uncle, a realist well-named "Snarl," growls, "O yes. He did not speak, but bleated his thanks to you." It is the uncle, too, who truly snarls what Shadwell must have considered a fitting epitaph upon not only Gimcrack, but all the virtuosi of the Royal Society: "If the blood of an ass were transfus'd into a virtuoso, you would not know the emittent ass from the recipient philosopher" (II.ii.212–15).

When Samuel Pepys left the theatre after he had seen *The Virtuoso*, which way, I wonder, did he turn? Did he join a group of fellow members of the Royal Society at one of their usual taverns? Or did he seek out "Dryden the poet," "Harris the player," and others at a "Witts' Coffee House?"

Appendix

Pepys,
Sir William Petty, and
the "Double Bottom"

T HE first mention of the "double Bottom" in Birch's *History of the Royal Society* was on November 12, 1662, when a letter of Petty's to Lord Brouncker "concerning his double-bottomed cylindrical vessel was read and ordered to be registered." Petty was "desired to prosecute this invention" and to report to the Society the success of the vessel at sea. On November 26 another letter of Petty's about the ship was read, with an extract from still another to John Graunt,[1] "Who was desired to let Sir William know that the Society was well pleased with the invention." Correspondence continued, with the result that in November, 1662, the Society appointed a committee in Ireland to examine the vessel and send a report, which was read at a meeting on January 28, 1663. I have earlier mentioned the prize offered by the Irish committee for a race, won by *Invention II*, carrying a flag indicating that it sailed under the aegis of the Royal Society. Whether because the committee was overexuberant or only

[1] Wheatley indexes separately "John Graunt" and the "Mr. Grant" from whom Pepys learned of the new boat. The names, however, refer to the same person, whose name was variously spelled "Grant" and "Graunt." In the roster of the Royal Society, to which he was elected on May 20, 1663, his name is given as "John Graunt." His election to the Society was presumably based upon his "Observations on a Collection of the London Bills of Mortality," a copy of which Pepys bought on March 24, 1662. This early and important study in statistics has frequently been attributed to Petty, who was closely associated with Graunt.

because the Society realized that this novel invention might well become important to the government, the Society informed its committee in Ireland on May 27, 1663, "that the matter of navigation being a state concern was not proper to be managed by the Society." [2] While Petty's name continued to appear in the minutes in various other connections, official interest of the Society in the "double Bottom" ceased.

I quoted in the last essay Pepys' description of the ship, based on letters shown him by Graunt. When Pepys first heard of it, *Invention II* had just won a wager of £50 on a race against the best ship His Majesty had in Irish waters. In his professional capacity as Clerk of the Navy, Pepys had every reason to follow with close interest all he could hear about this novel experiment in shipbuilding. References to it are frequent in 1663 and 1664. On August 5, 1663:

all the morning at the office, whither Deane of Woolwich came to me and discoursed of the body of ships, which I am now going to understand, and then I took him to the coffee-house, where he was very earnest against Mr. Grant's report in favour of Sir W. Petty's vessel, even to some passion on both sides almost.

One of several reasons for distrust among such shipbuilders as Anthony Deane [3] was that the "double Bottom" drew only one foot of water, which, while an advantage in shallow harbors, was held a disadvantage in deep water. This seems the basis for a remark made by the King that he "should be sorry it succeeded in Great, for then the Hollanders would have as much advantage of us as we do now of them. Because, as ships are now built, they can now have none so good as ours, but must draw more water than their Harbours would allow." [4]

[2] Quoted in *The Royal Society: Its Origins and Founders,* ed. Sir Harold Hartley (London, 1960), p. 85.

[3] Captain Anthony Deane, later Sir Anthony and Admiral, a well-known shipwright, had been appointed to Woolwich dockyard at the Restoration and became subsequently master shipwright at Harwich and Portsmouth. He had a distinguished political career as well.

[4] Quoted in Marquis of Lansdowne, ed., *The Double Bottom or Twin-hulled Ship of Sir William Petty* (Roxburghe Club; London, 1931), p. 64. This beauti-

In October, 1663, as I have pointed out, *Invention II* set out from Dublin to English waters. On October 14, 1663, Pepys noted: "So I toward the 'Change, and met with Mr. Grant, and he and I to the Coffee-house, where I understand by him that Sir W. Petty and his vessel are coming, and the King intends to go to Portsmouth to meet it." Let us return at this point to the anonymous "In Laudem Navis Geminae," [5] written while the vessel was in passage. The author was aware that Petty had originally built a model of the ship, which he had presented to the Royal Society, and that then he had built a small boat, *Invention I,* which we know from records was a vessel of one and three-quarters tons. The parodist wrote:

> Now Knight at last by deep Inspection
> Brings double Vessell to perfection,
> A little Gemini he built new,
> And did intend to have her gilt too,
> So strange a Thing was seen in no Age,
> For she to Rings-End made a Voyage.
> But doe not think Knight such an Elf
> To venture in Her first himself,
> For if he had, he had been doused,
> And in salt Sea like Mackrell Soused,
> For she (none knowing what's the matter)
> Turn'd up her Arse above the Water.
>   But this did nothing Knight dismay,
> Rome was not built up in a Day,
> A larger therefore he intended,
> And to have all her faultes amended,
> 'Cause her Bottoms hung down like Bollucks,
> He named her Castor and Pollux. [f. 76]

Thus, declares the balladmonger, we may trace the development of the boat that is attracting so much attention:

---

fully illustrated volume is the most authoritative study of the different models devised and built by Petty, consisting of letters and various other documents. Another of Petty's descendants, Lord Edmond Fitzmaurice, *The Life of Sir William Petty* (London, 1895), discusses the early vessels at some length (pp. 109 ff). The most recent study of Petty is by K. Theodore Hoppen in *History Today*, XV (1965), 126–34, with three illustrations.

[5] "In Laudem Navis Geminae E Portu Dublinii ad Regem Carolum II^dum Missae," Sloane MSS. 360, ff. 73–80, British Museum.

> From a little paper Frigot,
> Grew one of Wood, made of a Spigot,
> From thence a little bigger Toy,
> Which could but hold a Man, and Boy,
> Thence Castor, and Pollux, with double Bum,
> Doth to its full Perfection come.
> Thus from small Vessell in Tub swimming
> Is grown a Ship, with Sailes, and trimming,
> As big as other Ships, and able,
> To carry Masts, Anchor, and Cable
> And proudly sailing o're the Billows,
> Can send it faster than her Fellows. [f. 76v]

The unknown writer correctly distinguishes between *Invention I* and *Invention II,* just now setting out for England. Whoever the author, he knew a great deal about the circumstances under which the vessel set sail, though he apparently was not aware (at least did not mention) that Petty had built the ship with financial aid from Lord Massarene, who attempted to hold up the departure. He did know that Petty had had great difficulty in recruiting a crew, since both sailors and officers were not unnaturally reluctant to set out in a boat as experimental and as awkward as this one looked, with its two decks, and thirty tons:

> But Knight for Sea-Men being to seek,
> Was forc'd to use his Rhetorick,
> And with those Swabbers often jangle,
> 'Cause none would sail his Finglefangle:
> For at the pinch he found that Sailors,
> Had Hearts as little as any Tailors,
> And tho to see to lusty Swabbers,
> Were yet but Cowards, and fearful Lubbers.
> But at the last with Mony, and bawlings,
> He did pick up some few Tarpaulings,
> Who being valiant made by Brandy,
> Swore if he would they'd sayle to Candy. [f. 77]

The satirist was aware also that Petty's problems had not been limited to sailors, but that he had encountered great opposition from the wives of the officers and seamen he had finally procured. An account of Petty's powers of persuasion, as well as of his entertainment of the wives and children of his crew, is given in a letter from Sir Robert Southwell to Henry Oldenburg, dated October 13, 1663:

> This treatment Sir William gave his crew, their wives and children, on the night of embarquing; when having sent their wives and children in Ringsend coaches he provided them a banquet of burnt wine, stued prunes, applepyes, gingerbread, white sopps and milke, with apples and Nuts in abundance— and all this besides neates tongues and other more solid food for ye men themselves. And soe after much crying and laughing, hoping and fearing, he gott them all to part very quietly one from another, Intimating to ye women yt if they Succeeded there was a fleete to be built of double boddyed vessells, where-of every of their husbands would be a Capt; and in time it was not unlikely but they themselves would be laydes— Unto which they simperingly sayd, wiping their eyes, yt more unlikely things were come to passe.[6]

The satirist sets his corresponding scene of the descent of enraged maenads upon poor Petty somewhat earlier than the evening before the sailing, during the period when Petty was attempting to raise a crew:

> Their Wives heard on't (for they were maryed)
> And like mad Furies they came runing,
> From lousing of themselves, and suning,
> Swareing they should no strange Ship enter,
> Nor first goe on so great Adventure.
> Like Beldames then the Knight they threaten
> Who did expect he should be beaten. . . . [ff. 77–77v]

[6] Quoted by Lansdowne, pp. 91–92. The passage is also quoted in part by Fitzmaurice, p. 111. Fitzmaurice's editing of texts and his references are often exasperating, as may be evidenced in this instance: his footnote for this passage reads only, "Register Book of the Royal Society and Petty MSS." The Register Book is unpublished and the Petty MSS to which he refers are the Bowood Papers, privately owned by the Lansdowne family. The Marquis of Lansdowne also edited from these Bowood Papers the *Petty-Southwell Correspondence, 1676–1687* (London, 1928). For information in this note, I am indebted to the Berg Collection and Rare Books Division of the New York Public Library.

"Standing on tiptoes, And looking bigg without Muschatoes," the valiant knight ordered the bawling women to cease "their Furie, and their Catterwauling":

> With that the Knight took up his Station,
> And with loud voyce spake this Oration.
>   Fair Dames (quoth he) 'tis now or never,
> That all of you'd be made for ever,
> If you take what your Starrs doe offer,
> And not refuse now their kind proffer,
> For you that are Wives of poor saylors,
> As mean as wives of lousie Tailors,
> I doe foresee it, very shortly,
> You'l Gentlewomen be, and courtly.
> In truth I will declare what may be,
> Each one at least will be a Lady.
> Your linsie woolsie will be loathed,
> When you in Stuffs, and Silks are cloathed,
> Nor for your Raggs you'l not care a Louse,
> Nor any more to keep an Alehouse:
> Not wash your canns, nor wring the Spiggots
> When Husbands Captaines are of Frigots.[7] [f. 77v]

Not only will all their husbands be captains of men-of-war, but the first men to make a voyage in a double-bottomed boat will go down in history:

> The World won't for their Fames have stowage,
> Who first in double Ship made voyage,
> And all Men of them shall more speake,
> Than of Columbus, or Frank Drake.
> They shall be talk'd of on th' Exchange,
> For their voyage so wondrous strange,
> Their Names prick'd down in Gresham Colledg,
> For Men of Fame, of Skill, and Knowledg,
> In Brass ingrav'd with golden Letters.

[7] The satirist, concentrating on Petty's promise that the women will become "Gentlewomen," does not, of course, include some details which Southwell does in his letter: "Ye substance of ye speech which Sir William made to them was that he for himselfe was no sea-man, nor could tell what other vessells which they might Incounter with at Dublin were able to doe; but ye Intention was to send them with a vessell to his Maty, which though full of ugly faults and eye sores—being built for a fresh water Lough and to be carried 8 miles over land—was to outsayle any other vessell whatsoever and to endure all ye hazards of a troublesome passage from hence to London."

Then you'l take place before your Betters,
And every one shall have a Iewell,
As big, and broad as any Trouell.
And Thimbles, curling Pinns, and Bodkinns,
And Rings, and Braceletts, and such odd things,
And Pinns, and Gloves, and fine pin-cases,
And Knotts, Frizats, and long third lases,
With Forks, and silver handl'd Knives,
Fit presents for such worthy Wives,
That durst their Husbands first Adventure,
In double bottom'd Ship to enter,
Which our King sends for, for a tryall
To build by it a navie Royall. [f. 78]

Inevitably the satirist of the "double Bum" introduced into the purported speech of Sir William to the wives an argument not included in the Petty papers:

For on my word (beleev a stranger)
To sail my Ship there is no danger,
For there they'l rest, and sleep their fill,
And be as safe as Mouse in Mill.
For if one Bottom should grow leakie,
When Ship is tost, and Sailes are reakie,
And should too much of salt Sea quaff,
They may sail safe in to'ther ha'fe.
For so an ancient Poet sings,
Its good for Bows to have two Strings,
And as your selves may wisely note,
Two Oares belong still to one Boat,
Twould then concluded be at Gotham,
One Ship is best with double Bottom.
For if one side on rocks be staved,
They may on th'other still be saved,
It must indeed be wicked weather,
If both the Bottoms sink together. [f. 79]

In addition to all his other novelties, the ingenious Petty had presumably included in the structure of his remarkable vessel a device that prophesied a glorious future for the British navy, when the "double Bottom" had been adopted as the model for a first-class fighting ship:

Besides I'le have a new device,
Some Peggs untrussing in a trice,
The Ship that's one with double Bum,
Shall sever, and two Shipps become,
And be as like to one another,
As ever Sister was to Brother.
Will not this Project be a good one,
When Ships shall part thus on a sudden,
And fourty Sayle (which seem no more)
Shall presently become fowrescore.
Oh how wee'l meet the Dutch, and beat 'um
When we thus cuningly can cheat 'um. [ff. 78v, 79]

The oratory of the satirical knight, and the entertainment that accompanied it, met with as great success as had Sir William's, reported in the Petty MSS:

When they had heard what Knight had said,
They all were callm'd, their Wrath allay'd,
And every one incontinent
Gave Husbands leave with one Consent.
Then one of them among the rest,
A sooking Dame as it was ghess't,
Of good strong Liquor took a Quart,
Here's to you (quoth she) with all my heart,
I am resolv'd to whet my whistle,
And drink a good voyage to the Vessell,
That wee Sr Knight the better may thrive all
And she at London have safe arivall.
Strait to her Nose she setts the Stoop,
And supp'd it off as round as a hoop.
With that the Knight call'd for a whole Barrell,
Which put an end to all their Quarrell,
For he with Cupps ply'd them so roundly,
Till they were all Drunk most profoundly.
Thus he at last did Sailors gett,
With Mony, good Words, good Ale, and Wit. [ff. 79v 80]

If Petty saw these verses, he would not have been surprised. He had warned the crew that the new ship would provoke satire: [8]

[8] It is curious that although Petty's *Invention II* provoked satire in his own day, the Augustans failed to notice it in their satires on virtuosi of the

he advised them, if they did not believe she would answer these ends, they should not venture their lives to make them & him ridiculous. "For" sayd he "ye seamen & carpenters of London will have no mercy uppon you, nor will ye Court Witts & Poetts spare you. The Players will make Scenes of your double bodyes. Ye Citizens will make Pageants of yr Vessell uppon my Lord Mayor's day, but your comfort is that your King is to be your Judge. Youglets and Bizanios are your Enemyes—lett it not be sayd yt these hard outlandish sloopes should ever outcope you." [9]

The furies having been convinced to let their husbands go, Southwell describes the final exodus of the double-bottomed boat:

This being past, a douzen or 2 of beere more made flood enough to putt her under sayle. She went out with soe many boates hanging about her that made her look like a ship returned from ye East Indies overgrowne with great Barnacles; which having cast off and capps cast up, she made very fresh way.

Sir William and *Invention II* arrived in England some time in October, 1663. Petty's most important business during the next few months was to interest the King in his proposal to build a larger and speedier "double Bottom." Pepys' entries during the period indicate some of the opposition Petty encountered. Pepys mentioned a long conversation with Commissioner of the Navy Peter Pett on the same day Pepys had admired Petty's discourse on *Hudibras, Religio Medici,* and Osborne's *Advice to a Son,* January 27, 1664:

Restoration. Professor Kerby-Miller's statement of this is well made in his notes to the Scriblerian satire of Sir William's "pacing saddles": "How this invention [pacing saddles], an obscure failure of thirty years' standing, came to be mentioned in the *Memoirs* is puzzling. If the Scriblerians were trying to satirize Petty as an inventor they would have succeeded far better by referring to his famous failure, the 'double-bottom' or two-hulled ship, a model of which was kept in the museum of the Royal Society." See Charles Kerby-Miller, *Memoirs of Martinus Scriblerus* (New Haven, 1950), p. 334.

[9] Quoted Lansdowne, pp. 91–92. There is no way of knowing whether Southwell was quoting from memory, but his account is so similar to our satirist's that we may assume he is reporting to Oldenburg what he heard and saw fairly accurately.

He was mighty serious with me in discourse about the consequence of Sir W. Petty's boat, as the most dangerous thing in the world, if it should be practised by endangering our losse of the command of the seas and our trade, while the Turkes and others shall get the use of them, which, without doubt, by bearing more sayle will go faster than any other ships, and, not being of burden, our merchants cannot have the use of them and so will be at the mercy of their enemies.

Two days later, on January 29, "Mr. Deane" spoke against "Sir W. Petty's boat, which he says must needs prove a folly." Stubbornly loyal to his friend, Pepys commented, "I do not think so unless it be that the King will not have it encouraged." Pepys had every reason to anticipate discouragement two days later on February 1, 1664, when he heard the King laughing at Petty and Gresham College. On February 12, 1664, Pepys and Creed walked to Deptford, where Pepys met Petty: "I got him to go with me to his vessel and discourse it over to me, which he did very well." In some way Petty seems to have succeeded in his suit to the King, in spite of opposition. On December 22, 1664, as we have seen, both Pepys and Evelyn attended the launching of Petty's new "double Bottom," which the King christened *Experiment.* On February 13, 1665, Pepys went on board the *Experiment,* which seemed to him "a brave roomy vessel, and I hope may do well." A week later (February 18, 1665) Pepys was a member of the party invited to celebrate:

To the Royall Oake taverne in Lumbard Streete, where Sir William Petty and the owners of the double-bottomed boat (the Experiment) did entertain my Lord Brunkard, Sir R. Murrey, myself, and others, with marrow bones and a chine of beefe of the victuals they have made for this ship; and excellent company and good discourse; but, above all, I do value Sir W. Petty.

On March 22, 1665, at "Mr. Hubland's, the merchant," Pepys met Petty "and abundance of most ingenious men, owners and freighters of 'The Experiment,' now going with her two bodies to sea." *The Experiment* seems to have returned from that voyage and presumably from others. It was more than a year later that she was lost, with all hands, in a storm. As the satirist had said,

> It must indeed be wicked Weather
> If both the bottomes sink together.

The storm was such that, according to Anthony à Wood, seventy sail perished and no boat escaped. The fault, it would seem, was not in the ship but in its stars.

For nearly twenty years Petty made no further attempt to build another *Invention* or *Experiment*. About 1673-74 he devoted some attention to another possibility in shipbuilding that might have proved revolutionary: "An Attempt to demonstrate that an Engine may be fix'd in a good Ship of 5 or 600 Tons to give her fresh way at Sea in a Calm." So far as can be determined this scheme remained on paper.[10] In 1683-84, however, Petty returned to his great enthusiasm, in which Pepys took part in one way. "The fitte of the double bottom," Petty wrote to a friend, "do return very fiercely upon me. I cannot be dissuaded but that it contains most glorious, pleasant and useful things. My happiness lies in being mad." A paper among the Petty MSS shows a wager between Petty, on the one hand, and Pepys and Admiral Deane on the other: Sir William Petty's "15 Propositions touching his Sluce Boat," answered by Pepys and Deane. To each of Petty's boasts of what his ship would accomplish, the opposition replied in effect: "Shee shan't; for £500 upon each end." If Petty redeemed his wager, Pepys and Deane were winners of many hundreds of pounds on December 16, 1684, when in trial the new *Experiment* performed so abominably that even Petty was forced to acknowledge defeat. *Sic transit gloria navis geminae.*

---

[10] Fitzmaurice gives details of this project, pp. 122–24.

Page-Barbour Lecture Series

THE Page-Barbour Lecture Foundation was founded in 1907 by a gift from Mrs. Thomas Nelson Page (née Barbour) and the Honorable Thomas Nelson Page for the purpose of bringing to the University of Virginia each session a series of lectures by an eminent person in some field of scholarly endeavor. In a briefer form the materials in this volume were presented by Miss Marjorie Hope Nicolson in April, 1965, as the forty-eighth series sponsored by the Foundation.

Index

PEPYS' DIARY *AND THE NEW SCIENCE*

was composed, printed, and bound by
Vail-Ballou Press, Inc., Binghamton, New York.
The type is Baskerville,
and the paper is Warren's Olde Style.
Design is by Edward Foss.